T0140864

Object Shape Generation, Representation and Matching

DISSERTATION
zur Erlangung des Grades eines Doktors
der Ingenieurwissenschaften

vorgelegt von
M.Sc. Cong Yang
geb. am 10.02.1987 in China

eingereicht bei der Naturwissenschaftlich-Technischen Fakultät
der Universität Siegen
Siegen 2016

Bibliografische Information der Deutschen Nationalbibliothek

Die Deutsche Nationalbibliothek verzeichnet diese Publikation in der Deutschen Nationalbibliografie; detaillierte bibliografische Daten sind im Internet über http://dnb.d-nb.de abrufbar.

ISBN 978-3-8325-4399-0

Logos Verlag Berlin GmbH
Comeniushof, Gubener Str. 47,
10243 Berlin
Tel.: +49 (0)30 42 85 10 90
Fax: +49 (0)30 42 85 10 92
INTERNET: http://www.logos-verlag.de

Acknowledgements

First of all, I would like to thank Prof. Dr. Marcin Grzegorzek, Prof. Dr. Udo Kelter, Prof. Dr. Andreas Kolb and Prof. Dr. Dietrich Paulus, for serving on my committee. I appreciate the time they set aside for me and am thankful for the invaluable discussions and comments regarding this work.

I would like to express my gratitude to my supervisor, Prof. Dr. Marcin Grzegorzek, for his incredible supports and countless advices that finally lead to the success of this work. Prof. Marcin has always taken the time to give me advices and share his good ideas on my research problems. Working with him, I have learned not only how to think through a problem but also how to express my ideas in a precise and concise way. I am eternally grateful for his guidance and patience with me along the way.

I want to extend a very special thanks to Dr. Kimiaki Shirahama for proofreading my papers and helping me with many research ideas. The benefit of these experiences is invaluable. Moreover, I want to express my appreciation to Dr. Christian Feinen and Oliver Tiebe for their invaluable helps during my staying in Siegen as well as their cooperation on my research.

Also, a big thanks to Zeyd Boukhers, Lukas Köping, Ahmad Delforouzi, Muhammad Hassan Khan, Kristin Klaas, Dr. Amir Tabatabaei, Dr. Hamed Shariat Yazdi and Marcel Piotraschke, etc. for being great colleagues and friends. I am extremely happy to have spent the past four years working with them. In addition, a special thanks to Dr. Chunyang Chen and Dr. Fang Yang for their great help, encourage and accompany in Germany.

Over the years I have been fortunate to have successful collaborations with multiple researchers in different institutes: Dr. Ewa Łukasik (Poznan University of Technology), Prof. Dr. Bipin Indurkhya (Jagiellonian University), Dr. Olaf Flak (University of Silesia), Pit Pietsch and Dominik Scarpin (University of Siegen). Thank you for bringing me into new areas. Additionally, a big thanks to Ewa and Bipin for inviting and helping me during my visit in Poznan and Krakow, respectively. I had a great and memorable time in Poland!

Finally, and most of all, I would like to express my love and gratitude to my parents, grandparents and my sister for being with me and helping me over many difficulties.

Abstract

Shape matching and recognition is a fundamental aspect of many problems in computer vision, including object or scene recognition, moving tracking and object detection, etc. While rigid shape matching is relatively well understood, matching shapes undergoing non-rigid deformations remains challenging. Moreover, shape generation is also a challenging task since nearly all approaches face the same difficulty: background clutter. The aim of the research in this thesis is to present novel approaches to generate, represent and match object shapes. To achieve this, the following three problems are addressed in particular:
Shape Generation: Shape generation is generally applied based on shape contour detection which is a process to locate an object, identify its contour parts, and segment out its contour. After that, an object shape can be generated by properly filling hollow spaces. In order to detect shape contour parts (open curves), a Contour Segment (CS)-based open curve matching approach is introduced. CS is the basic element for representing and matching open curves. However, there exist various CS descriptors and the discrimination power of each descriptor is unclear. Furthermore, deep insights are needed regarding the suitable combination between CS descriptors and their correlated matching methods. Therefore, this thesis studies and evaluates 26 CS descriptors in a structured way. Formed on evaluations, the recommended CS descriptors and their correlated matching algorithms are employed for open curve matching.
Shape Representation: Shape-based object representation looks for effective and perceptually important shape features based on either shape boundary or region information. However, there are three main challenges. The first one is how to extract efficient descriptors that are invariant to shape rotation, translation and scaling. The second one is how to extract shape descriptors that are robust to noise and distortions. The third challenge is how to generate descriptors with low computational complexity.

In order to solve these problems, in this thesis, one way is to propose simple geometry descriptors (coarse-grained descriptors) for capturing coarse-grained shape features and then fuse them with existing descriptors to improve the description power of individuals. Another way is to propose rich descriptors (fine-grained descriptors) which have a higher description power since the "detailed" shape features are captured. Experiments show that the proposed coarse-grained shape descriptors can significantly improve the description power of existing descriptors without adding too much computational complexity. Furthermore, the proposed fine-grained descriptors achieves the state-of-the-art performances in several standard datasets.

Shape Matching: Based on descriptors, shape matching aims to calculate the overall similarity (or dissimilarity) between two object shapes. There is no general model for shape matching since different descriptors may have different feature vector structures. In addition, even with the same structure, a matching method is required by considering both accuracy and speed. Specifically, for coarse-grained shape descriptors, the matching between shapes is usually conducted by using simple vector distance functions. Thus, the matching accuracy could be improved by optimising distance functions without increasing the computational complexity. In contrast, fine-grained shape descriptors have complex structures and the matching process is normally applied based on correspondences and statistics. Therefore, the matching accuracy could be improved by exploring extra information from descriptors to enhance the correspondences and statistics.

With these observations, several matching algorithms are proposed in this thesis. Specifically, for coarse-grained descriptors, a supervised optimisation strategy is proposed to improve the flexibility and adaptability of their feature vectors. For fine-grained descriptors, inspired by the idea of high-order graph matching, the singleton, pairwise and third-order potential functions are proposed to solve the interesting point and hierarchical skeleton matching problems. In addition to the coarse- and fine-grained shape descriptors, three integrated descriptors are also delivered and evaluated.

Lastly, two shape-based applications are introduced and assessed using the proposed descriptors: (1) With a coarse-grained descriptor, an automatic system for classifying Environmental Microorganism (EM) is introduced. Their classification is a very important indicator for biological treatment processes and environmental quality evaluations. Experimental results certify the effectiveness and practicability of the proposed EM classification system as well as the shape-based method. (2) With a fine-grained descriptor, a novel skeleton-based approach is proposed for audio envelope analysis. To effectively describe the envelope of audio signals, a skeleton-based descriptor and its matching method are introduced. The application of violin sound matching proves the usability of these approaches.

Zusammenfassung

Die Form ist eine expressive Abstraktion des visuellen Musters eines Objekts. Formabgleich und Formerkennung sind fundamentale Aspekte des maschinellen Sehens, insbesondere der Objekt- und Szenenerkennung, der Bewegungserfassung und Objektdetektion, etc. Während der Abgleich starrer Formen bereits relativ gut verstanden wird, bleibt der Abgleich nicht-starrer, deformierter Formen herausfordernd. Des Weiteren ist auch die Formgenerierung eine herausfordernde Aufgabe, da sämtliche Ansätze dem Problem eines starken Hintergrundrauschens begegnen. Die in dieser Dissertation präsentierte Forschung zeigt neuartige Ansätze zur Generierung, Repräsentation und zum Abgleich von Objektformen auf. Um dies zu erreichen, werden speziell die drei folgenden Probleme adressiert:

Formgenerierung: die Formgenerierung basiert auf der Konturdetektion, welche ein Objekt lokalisiert, die Konturbestandteile identifiziert und die Segmentierung vornimmt. Anschließend kann die Form durch Ausfüllen der innen liegenden Flächen generiert werden. Zur Detektion der Konturbestandteile wird der Ansatz eines kontursegmentbasierten Abgleichs offener Kurven eingeführt. Ein Kontursegment ist das Basiselement zur Repräsentation und zum Abgleich offener Kurven. Allerdings existieren verschiedene Deskriptoren für Kontursegmente, wobei deren Leistungsfähigkeit zur Unterscheidung von Formen unklar ist. Des Weiteren sind für eine geeignete Kombination verschiedener Kontursegmentdeskriptoren und der entsprechenden Abgleichungsalgorithmen tiefgreifende Erkenntnisse von Nöten. Daher werden in dieser Dissertation 26 Kontursegmentdeskriptoren strukturiert untersucht und evaluiert. Die nach der Evaluierung empfehlenswerten Deskriptoren und die entsprechenden Abgleichungsalgorithmen werden für den Abgleich offener Kurven angewandt.

Formrepräsentation: Methoden der formbasierten Objektrepräsentation suchen nach effektiven und für die Wahrnehmung bedeutsamen Formmerkmalen, welche entweder auf Konturen oder regionalen Informationen basieren. Allerdings gibt es dabei drei Herausforderungen. Die Erste ist das Extrahieren effizienter Deskriptoren, welche invariant bezüglich Rotation, Translation und Skalierung sind. Die Zweite ist das Extrahieren von Deskriptoren, die robust bezüglich Rauschen und Verzerrung sind. Die dritte Herausforderung besteht darin, Deskriptoren mit einer geringen Berechnungskomplexität zu generieren.

Diese Dissertation löst diese Probleme, indem einerseits einfache geometrische Deskriptoren zur Gewinnung grobgranularer Formmerkmale vorgestellt werden, welche mit existierenden Deskriptoren verknüpft werden, um deren Fähigkeit zur Unterscheidung von Formen zu verbessern. Andererseits werden feingranulare Deskriptoren, welche aufgrund ihres Detailgrades eine höhere Unterscheidungsfähigkeit besitzen, verwendet, um detaillierte Formmerkmale zu extrahieren. Experimente zeigen, dass die vorgestellten grobgranularen Formmerkmale die Unterscheidungsfähigkeit bekannter Deskriptoren signifikant steigern können, ohne den Berechnungsaufwand bedeutend zu erhöhen. Des Weiteren erreichen die vorgestellten feingranularen Deskriptoren in Standarddatensätzen Leistungen des aktuellen Forschungsstands.

Formabgleich: Anhand der Desktriptoren berechnet der Formabgleich die Ähnlichkeit zweier Objektformen. Es gibt kein allgemeingültiges Modell für den Objektabgleich, da verschiedene Deskriptoren unterschiedliche Merkmalsstrukturen haben können. Des Weiteren muss bei der Wahl der Abgleichsmethode selbst bei gleicher Merkmalsstruktur Genauigkeit und Laufzeit berücksichtigt werden. Für grobgranulare Deskriptoren werden einfache Vektordistanzfunktionen zum Abgleich verwendet. Folglich kann die Genauigkeit durch Optimierungen der Distanzfunktionen erhöht werden ohne dabei die Komplexität zu erhöhen. Im Kontrast dazu haben feingranulare Deskriptoren sehr detaillierte Strukturen, sodass der Abgleichprozess anhand von Äquivalentenfindung und Statistiken durchgeführt wird. Daher könnte die Abgleichgenauigkeit durch das Entdecken weiterer Informationen über das Verhalten der Deskriptoren zur Verbesserung der Äquivalentenfindung und Statistiken erhöht werden.

Mit diesen Beobachtungen werden einige Abgleichungsalgorithmen in dieser Dissertation vorgestellt. Speziell für grobgranulare Deskriptoren wird eine überwachte Optimierungsstrategie vorgestellt, und die Flexibilität und Adaptivität der Merkmalsvektoren verbessert. Für feingranulare Deskriptoren wird ein Ansatz vorgestellt, welcher durch das High-order Graph Matching inspiriert wurde. Mittels Potentialfunktionen einfacher, paarweiser sowie dritter Ordnung soll sowohl das Problem des Abgleichs von Punkten, als auch das Problem des Abgleichs hierarchischer Skelettstrukturen gelöst werden. Zusätzlich zu den grob- und feingranularen Deskriptoren werden außerdem drei integrierte Deskriptoren evaluiert.

Abschließend werden zwei formbasierte Anwendungen vorgestellt und anhand der vorgestellten Deskriptoren bewertet: (1) Mittels grobgranularer Deskriptoren wird ein automatisches System zur Klassifizierung von Mikroorganismen vorgestellt. Die Klassifizierung von Mikroorganismen ist ein wichtiger Indikator in biologischen Betrachtungen sowie Evaluierungen der Qualität unserer Umwelt. Experimente weisen die Effektivität und Umsetzbarkeit des vorgestellten Systems sowie des formbasierten Ansatzes nach. (2) Mit einem feingranularen Deskriptor wird ein neuartiger, skelettbasierter Ansatz zur Audiospektrumsanalyse vorgestellt. Um das Spektrum von Audiosignalen effektiv zu beschreiben, werden ein skelettbasierter Deskriptor und eine dazu passende Abgleichungsmethode eingeführt. Die Anwendung des Abgleichs von Violinenklängen beweist die Verwendbarkeit solcher Ansätze.

Contents

Chapter 1

Introduction

Shape is an expressive abstraction of the visual pattern of an object (see Figure 1.1). For human beings, shape is a noticeable attribute of the world around us and our perception system uses it as a primary feature to detect and identify objects [SBC08]. For example, in most early childhood programmes, object shapes are taught early in the year rather than letters and numbers [Chu14]. In the domain of computer vision, shape is being used in many applications [YIJ08; YT07; Wan+11; Ren+13; WL04; CT13] to recognise objects in images.

In order to adapt to the demands of different scenarios, proper shape generation, representation and matching algorithms are required. Therefore, this thesis addresses these requirements in a structured way. Firstly, open curve-based shape contour detection methods are discussed and evaluated for shape generation. Secondly, several efficient coarse- and fine-grained shape descriptors are proposed for shape representation in different scenarios. Fine-grained descriptors feature the small shape deformation while coarse-grained descriptors capture the global shape deformations. Thirdly, for each descriptor, a matching algorithm is designed by considering both accuracy and speed. Finally, the usability of the proposed methods is validated by experiments and real-world applications.

Figure 1.1: An object (crocodile) in the image[1] and its shape.

[1] http://toutiao.com/a4715232748/, [online: 14th July 2015]

1.1 Problems and Overall Goals

Most shape-based applications [YIJ08] face the same challenge: object deformation. As shown in Figure 1.2, the shapes of the same object are visually different depending on its deformations and perspective. Moreover, an image with large uncertainties in segmentation due to background clutter could also lead to various object shapes. To overcome this, the problems are discussed and worked out within three aspects: (1) Shape generation, (2) shape representation and (3) shape matching.

Figure 1.2: An illustration of shapes which significantly vary depending on deformations.

1.1.1 Shape Generation

Shape generation allows the localisation and segmentation of previously unseen objects in images. However, in many cases, an object shape cannot be ideally segmented due to the background clutter [Har+15] and object overlaps [Liu+13]. In order to reduce uncertain segmentations, two main approaches can be employed [RDB10a]: appearance and contour. Appearance-based approaches normally form the conspicuous objects in an image using the bag-of-words model [SZ03] which analyses the orderless distribution of local image features. However, these approaches highly depend on the local image descriptors. Moreover, they are not robust to objects with multiple patterns (colours and textures) [MS05].

Contour-based approaches detect and segment object shapes based on open curve matching and grouping [BST15]. Open curve is a fragment of a shape boundary and it provides powerful and often more generic features. Since the connectedness of shape boundary points is not ensured in practice due to the noise affections, viewing open curves as local patches with any length provides more flexibility against boundary instabilities. Thus, compared to appearance-based approaches, contour-based approaches are more robust to large variations in texture or colour. In order to efficiently match open curves, many different approaches have been proposed [YIJ08; RDB10a]. However, the discrimination power of each descriptor is unclear, and how similar or different it is to the other descriptors. Furthermore, deep insights are needed regarding the suitable combination between representation and matching algorithms. The overall goal of this thesis part is to evaluate different open curve representation and matching methods. The evaluated results are then used in the framework of shape contour detection.

1.1.2 Shape-based Object Representation

Shape-based object representation generally looks for effective and perceptually important shape features based on either shape boundary information or boundary plus interior content [ZL04]. However, there are three main challenges. The first one is how to extract efficient descriptors that are invariant to shape rotation, translation and scaling. The second one is how to extract shape descriptors that are robust to noise and distortions. This is known as the robustness requirement. The third challenge is how to generate descriptors with a low computational complexity.

In order to solve these problems, one possible way is to use the simple geometry descriptors such as [ZL04; Yan+15b] for capturing coarse-grained shape features. They have low computational complexity and most of them are robust to noise and distortions. However, these descriptors have limited description power because they are too simple and lack of information [ZL04]. Another possibility is to use some rich descriptors [BL08; BMP02; HS15; LJ07] which have a higher description power since the fine-grained shape features are captured. Although most of them have better invariance properties, high computational costs for the feature extraction and matching are required [HS15]. Thus, the overall goal of this part is to propose efficient descriptors in both fine- and coarse-grained levels for capturing geometrical and topological shape features.

1.1.3 Shape-based Object Matching

Based on descriptors, shape-based object matching aims to calculate the overall similarity (or dissimilarity) between two object shapes. There is no general model for shape matching since different descriptors may have different feature vector structures [Yan+15b; BMP02]. Moreover, even with the same structure, a matching method is required considering both accuracy and speed. Therefore, the main question of shape matching is how to improve the matching accuracy without adding too much computational complexity.

Specifically, for coarse-grained shape descriptors, the matching between shapes is a straightforward process, which is usually conducted by simple distance functions [Yan+15b; Yan+14b]. Thus, the matching accuracy could be improved by optimising this class of functions without increasing the computational complexity. In contrast, fine-grained shape descriptors have complex structures and the matching process is normally applied based on correspondences [BL08; BMP02] and statistics [MSJB15; Wan+12a]. Therefore, the matching accuracy could be improved by exploring extra information from descriptors to enhance the correspondences and statistics. With these observations, the overall goal of this part is to propose proper matching algorithms for different types of shape descriptors.

1.2 Overview and Contributions

The next four chapters address the detailed approaches for shape generation, representation and matching. Specifically, in Chapter 2, the related shape generation methods are firstly introduced; then, different open curve representation and matching algorithms are surveyed and evaluated in the context of open curve-based shape contour detection. The open curve evaluation work is based on [Yan+16a] and also contributes three new annotated datasets to the community. The next three chapters detail the main contributions of this thesis:

- In **Chapter 3**, shape representation with coarse-grained features is discussed. This chapter presents two simple and intuitive shape representation methods with shape boundary partitions and bounding boxes. These descriptors can effectively capture the coarse-grained features for shape matching. These works were published in [Yan+14b; Yan+15b].
- In **Chapter 4**, shape representation with fine-grained features is addressed. In this chapter, two rich shape descriptors are proposed where one uses the contour-based method and the another one employs the region-based method. These descriptors are robust to different shape deformations and achieve high matching accuracy in the experiments. These works were published in [Yan+16d; Yan+16b; Yan+16c; Hed+13].
- In **Chapter 5**, matching algorithms for the aforementioned shape descriptors are introduced. Moreover, this part also presents the matching algorithms using the integrated coarse- and fine-grained features. These matching algorithms are not only designed for the proposed shape descriptors, but also can be built on other existing shape descriptors. These works were published in [Yan+14b; Yan+15b; Yan+16d; Yan+16b; Yan+16c; Hed+13; Fei+14; YG14].

The remaining chapters present the experimental results and applications of the proposed approaches. More specifically, Chapter 6 gathers all results which have been monitored during the evaluation of each shape descriptor and its correlated matching algorithm. The performance of the shape generation is also presented in this chapter. In Chapter 7, some applications using the proposed coarse- and fine-grained descriptors are introduced and evaluated. Finally, conclusions are drawn in Chapter 8 by summarising all observations during the experiments. Moreover, some further possible directions and extensions are discussed.

The most significant scientific contributions of this thesis are: (1) For shape generation, the existing open curve representation and matching approaches are surveyed and evaluated in a structured way. Formed on evaluations, the combinations of representation and matching algorithms are recommended for the shape contour detection. (2) Two simple and intuitive descriptors are proposed for capturing coarse-grained shape features. These descriptors can be used for fast shape matching and retrieval with promising accuracy. Moreover, they can also be fused with rich shape descriptors to improve their discriminating power. (3) Two rich descriptors are introduced for preserving fine-grained shape features. These descriptors are used for accurate shape matching and retrieval. Compared to existing rich descriptors, the proposed methods achieve a high performance without adding too much computational complexity. Furthermore, a proper integration of coarse- and fine-grained descriptors can improve the matching accuracy over the original one. (4) For each proposed descriptor, a correlated matching algorithm is properly designed. (5) It is shown that extensions of the proposed shape-based approaches can be used for some real-world applications.

Chapter 2

Shape Contour Detection Based on Open Curves

Shape contour detection is a process to locate an object, identify its contour parts, and segment out its contour. After that, an object shape can be generated by properly filling hollow spaces. While there are many different methods designed for shape contour detection, there is one difficulty faced by nearly all approaches: Background clutter [Lin+15]. In this thesis, shape contours are detected by open curving matching. In general, an open curve can be defined as a curve with two endpoints (e.g. Figure 2.1 (a)). Comparing to a close curve (e.g. Figure 2.1 (b)) with no endpoint, the open curve endpoints do not join up. An intuitive understanding is that a close curve is a shape contour and an open curve is a shape contour fragment with any length. In some literatures [Can86; HS92; SBC08], open curve is also called edge or contour fragment.

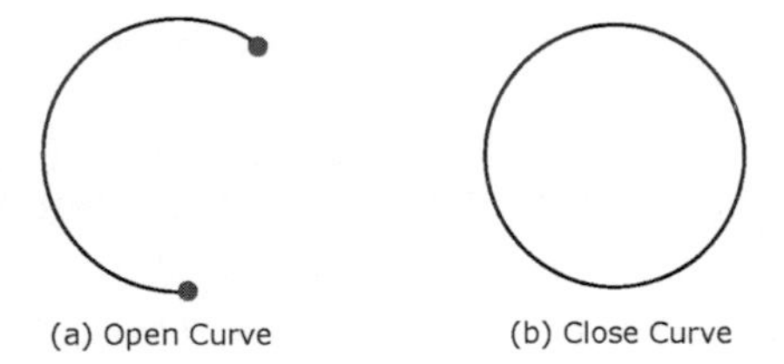

Figure 2.1: Examples of an open and a close curve. The open curve endpoints are marked by the red colour.

Shape contour detection by open curve matching can properly handle the background clutter problem. The main reason is that both shape contour and background information can be preserved and distinguished by open curves. Specifically, given an image (Figure 2.2), an open curve map is firstly generated with object boundaries and background edges using intensity gradients [Can86] and zero-crossings [HS92] approaches. After that, an open curve reduction process is applied using open curve dropping [Zha+07; AL08] and linking [DRB10a] methods. With the above mentioned processes, an image is represented by various open curves (open curve book) with different lengths and deformations. Formed on open curves, the shape contour detection task is addressed by solving the contour vs non-contour classification problem [KWH14; MYP14]. Finally, some further tasks like object localisation and shape generation can be applied based on the shape contour detection results. In this chapter, the detailed processes of open curve-based shape contour detection are introduced. Moreover, different open curve representation approaches are evaluated for open curve matching and shape contour detection. Please notice that although open-curved based shape contour detection methods can generate shapes with labelled classes, the shape representation and matching processes introduced in Chapter 3~5 are still necessary. The main reason is that in many applications [BL08; Yan+14a], similarities between shapes are required for searching similar objects in a database. In this case, shape labels cannot be used for accurate object retrieval.

2.1 Techniques for Shape Contour Detection

There are several works that use the open curve-based approach for shape contour detection. [FJS09] gives a brief overview of some related methods. Based on an open curve book, those methods use a voting of configurations of open curves followed by a refinement scheme to discover shape from example models (e.g. the goose model in Figure 2.2). Specifically, [SZS10] introduces an approach in which long, salient and image open curves are employed to learn and detect object shapes. Different from traditional methods that use one-to-one matching of open curves to a model, they propose a many-to-one matching scheme since image open curves often fragment unpredictably. In [Lu+09a], the authors use a particle filter method for recognising and grouping open curves with a given model. Particularly, they extend the standard particle filter framework to drive an intuitive search strategy for the target contours in an open curve book. In such a case, particles in their approach perform tracking in the space of pairs of model contour fragments and the open curves. However, this method expects open curves to have consistent fragmentation across images, which is not always satisfied. In order to solve this problem, in [RDB10a], a particle open curve matching method is proposed for shape contour detection. Recently, driven by the prevailing trend of deep learning methods [LBH15], the deep neural networks [AGM14] are used for object shape detection and have achieved impressive performances [BST15; She+15].

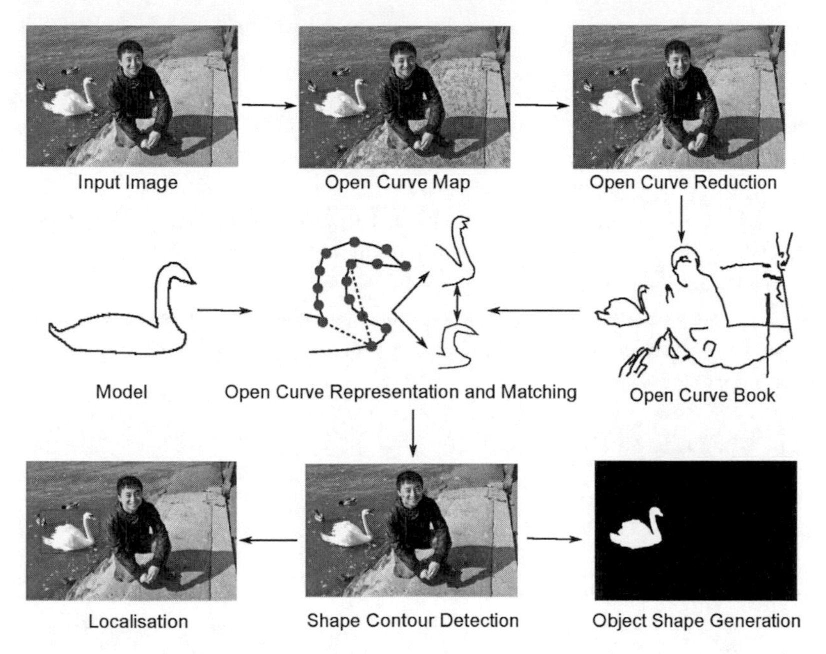

Figure 2.2: Pipeline of shape contour detection and shape generation using open curves.

Instead of detecting contours with low-level cues for a higher-level recognition task, these methods exploit object-related features as high-level cues for contour detection. With this, the contour detection accuracy can be improved by making the use of the deep features learned from Convolutional Neural Networks (CNNs). For each approach introduced above, a proper open curve representation method is required. However, most existing works only introduce open curve representation approaches without any comparison. In addition, some survey papers focus on shape representation techniques [ZL04; YIJ08], or only compare a handful of open curve representation in specific applications [PKB11]. There is no paper that integrally surveys and evaluates open curve representation techniques. Thus, in this chapter, in addition to the detailed introduction of the open curve-based shape contour detection method, the open curve representation methods are also surveyed and evaluated.

2.2 Open Curve Generation from Images

Open curve generation[1] methods identify points in a digital image at which the image brightness changes sharply or, more formally, has discontinuities. The points at which image brightness changes sharply are typically organised into a set of open curves [MH80]. The purpose of detecting sharp changes in image brightness is to capture important events and changes in the properties of the world. In the first part of this section, the detailed process for detecting shape changes in an image is introduced. However, due to the affections of blurs and shadings on a natural image, the obtained open curve map normally contains a mass of noise which seriously impact the contour detection accuracy and speed. In order to remove noises from the generated open curve map, in the second part, an open curve reduction method is illustrated.

2.2.1 Open Curve Generation

There are many methods [ZT98] for open curve generation and a general survey can be found in [BM12]. Here, the Canny method [Can86] is employed and introduced since it is still a state-of-the-art method and adaptable to various environments. Following are the brief description of its stages:

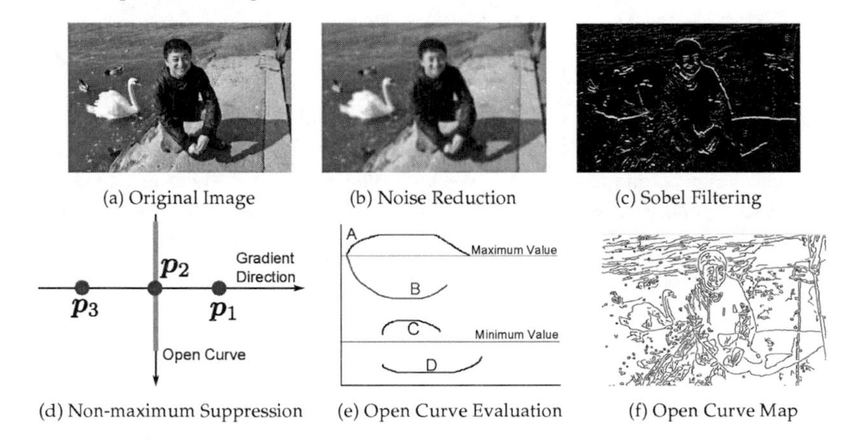

Figure 2.3: Open curve generation from an image with the Canny method [Can86].

[1] In some literatures, it is called Edge Detection.

- (1) For a given image, the first step is to remove noises in the image with the Gaussian filter [YVV95] since open curves are susceptible to noise in the image (Figure 2.3 (b)).
- (2) The smoothed image is then filtered with a Sobel kernel [BT88] in both horizontal and vertical directions so that the gradient and direction for each pixel can be observed (Figure 2.3 (c)).
- (3) After obtaining gradient magnitude and direction, a full scan of the image is done to remove any unwanted pixels which may not constitute open curves. For this, each pixel is checked in terms of whether it is a local maximum in its neighbourhood in the direction of gradient. For example, in Figure 2.3 (d), a point $\mathbf{p}_2$ is on an open curve in vertical direction and its gradient direction is normal. Point $\mathbf{p}_1$ and $\mathbf{p}_3$ are in gradient directions so that point $\mathbf{p}_2$ is checked with point $\mathbf{p}_1$ and $\mathbf{p}_3$ to see if it forms a local maximum. If so, it is considered for the next stage. Otherwise, it is suppressed.
- (4) In the last stage, the generated curves are evaluated and decided for the final output. The purpose of this stage is to remove small noisy pixels on the assumption that open curves are long lines. For this, two threshold values are defined: Minimal and maximum values. If the intensity gradient of an open curve is more than the maximum value, it will be kept for the output (e.g. the open curve A in Figure 2.3 (e)). If one is below the minimal value, it will be discarded (e.g. the open curve D in Figure 2.3 (e)). For an open curve lying between these two values, an additional evaluation is applied. Specifically, if it connects to an output open curve, it will be considered to be part of this open curve. Otherwise, it will also be discarded. For instance, in Figure 2.3 (e), although open curve B is below the maximum value, it is considered as a valid open curve since it connects to the open curve A. But open curve C is discarded since it is isolated.

Based on the previous stages, as shown in Figure 2.3 (f), an open curve map is generated for the final output. However, it can be observed that even with a reduction process in the last stage, there still remain a plenty of small open curves. The main reason is that the employed maximum and minimum values are not general for every images. For shape contour detection, those small open curves could seriously influence the detection accuracy and time. Thus, the second round of open curve reduction is required.

2.2.2 Open Curve Reduction

In this round, an open curve map is simplified by two evaluation steps: (1) length evaluation and (2) stability evaluation. Specifically, in the first step, a threshold t_{length} is defined for removing the small open curves. If the length of an open curve is less than t_{length}, it will be discarded. Otherwise, it will be kept for the next step. The motivation of this step is built on the fact that the interest open curves normally have a higher connectivity [DRB10a]. In practice, t_{length} is an empirical value which is defined by considering the resolution and curve statistics of an open curve map.

In the second step, the stability of each open curve is evaluated with the method in [DRB10a]. This method evaluates open curves by analysing regions that support the local gradient magnitudes. Particularly, a component tree is built for encoding the gradient magnitude structure of the input image. In contrast to common curve evaluation algorithms [Can86] which only analyse local discontinuities in image brightness, a component tree integrates mid-level information by analysing regions that support the local gradient magnitudes. In a component tree, each node contains a single connected region obtained from the gradient magnitude image with different thresholds. Tree edges are defined by an inclusion relationship between nested regions in different levels of the tree. Based on the component tree, the stability is estimated by comparing region sizes between nodes in adjacent levels of the tree. After calculating the stability values for every node in the component tree, the most stable ones are detected and their correlated open curves are returned.

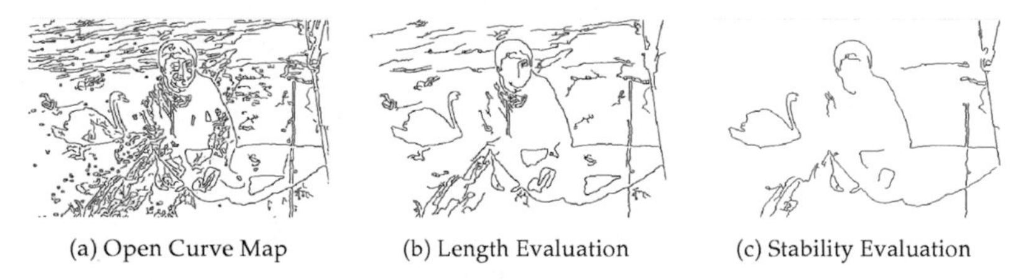

(a) Open Curve Map (b) Length Evaluation (c) Stability Evaluation

Figure 2.4: Open curve reduction based on the length and stability evaluation steps.

In Figure 2.4, for a given open curve map in (a), the simplified maps based on length and stability evaluations are illustrated in (b) and (c), respectively. It can be clearly observed that most small open curves are removed. Finally, the remaining open curves in (c) are used for constructing the open curve book (Figure 2.2) for shape contour detection. In practice, the second round open curve reduction is optional since the sizes of target objects are varying in different images. If an object's size is smaller than the thresholds (like the grey ducks in Figure 2.2 (a)), it could be removed from the open curve book. Thus, the second round reduction is applied based on different scenarios.

2.3 Open Curve Matching using Contour Segments

Based on the open curve book, shape contour detection is applied built on the matching between the partial model contours and open curves. Specifically, as shown in Figure 2.2, each open curve in an open curve book is compared to an open curve from the contour model. Open curve matching aims to find similar parts between two open curves and then calculate their similarity.

In this thesis, open curves are represented and matched based on Contour Segment (CS). A CS is a fragment of an open curve which is constructed by a chain of connected open curve points. As shown in Figure 2.5, compared to an open curve which is defined as a sequence of curve points, a CS only describes the partial information of an open curve. Each curve point in the CS is called a CS point. The main motivation for CS is that the lengths and deformations of generated open curves in an open curve book are not uniformed. Thus, viewing CSs as local patches with any length provides more flexibility for open curve representation and matching.

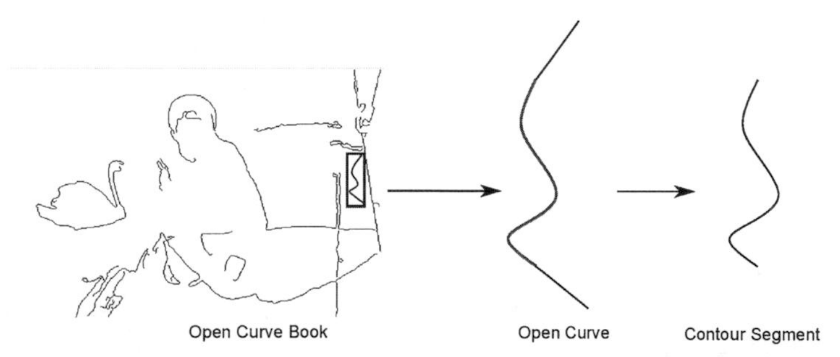

Figure 2.5: Open curve and a contour segment (marked with the red line).

Benefiting from these properties, CSs are used to represent the partial features of an open curve, then the similar parts between open curves are searched by CS matching. In order to efficiently match CSs, proper CS descriptors and matching algorithms are required. Such a descriptor is used to encode the geometric configuration of those CS points. More specifically, among all CS points, sample points are normally selected to generate CS descriptors by considering the geometrical relationship between those points. If sample points are selected roughly, CS descriptors represent the coarse-grained CS features. If sample points are selected densely, CS descriptors describe the fine-grained CS features. For CS matching, traditional matching algorithms like Hungarian [Kuh55], Dynamic Programming (DP) [Ric54] or Dynamic Time Warping (DTW) [ACT09] are normally employed for searching the correspondences between sample points. For some CS descriptors [MSJB15; Yan+14b], due to the structure of their feature vectors, distance functions like correlation [YK8], Histogram Intersection (HI) [RTG00], χ^2-statistics [RTG00] and Hellinger [Bha46] are used for calculating their distances.

Despite the extensive usage of CS [RDB10a; Yan+14b; WKL15; Mah+14] and its extensions [ML11a; DRB10b; MSJB15], most existing works only introduce CS descriptors without any comparison. In addition, some survey papers focus on shape representation techniques [ZL04; YIJ08], or only compare a handful of CS descriptors in specific applications [PKB11]. There is no related work that integrally surveys and evaluates CS descriptors. In other words, the discrimination power of each descriptor is unclear; as well as how similar or different it is to the other descriptors. Furthermore, deep insights are needed regarding the suitable combination between CS descriptors and matching algorithms, and the computational efficiency of each descriptor. Therefore, this section studies the invariance properties, matching performance and computation complexity of 26 CS descriptors and their matching algorithms in a structured way. Formed on evaluations, the combinations of CS descriptors and matching algorithms are recommended for open curve matching with different requirements in terms of accuracy and speed.

2.3.1 Contour Segment Descriptors

In this section, CS descriptors are put into three groups, "simple", "signature-based" and "rich". Here, CS descriptors are classified and grouped based on the structure of their feature vectors. In order to regulate and standardise the next descriptions, for a CS, the following restrictions are addressed: (1) With one pixel width. (2) With two endpoints, which only have one neighbour point. (3) No intersection point: except two endpoints, the rest of points only have two neighbour points. Formed on these restrictions, a CS C is represented as a sequence of CS points $\mathbf{p}_1, \mathbf{p}_2, \cdots, \mathbf{p}_N$ along the curve path. Here, a CS point $\mathbf{p}_i (1 \leqslant i \leqslant N)$ is expressed as a point in the Cartesian coordinate system, that is, $[x_i, y_i]$. CSs are set to have the same number of points for two reasons: First, it is easier to fairly evaluate the performance of CS descriptors using the same matching methods. Second, it also makes open curve matching easier since open curves can be decomposed into multiple CSs with the same number of points. Then, an open curve matching task is accomplished by multiple CS matching tasks.

In order to reduce the time complexity for CS matching, in practice, rather than the full CS points, only part of the CS points are selected as the sample points for calculating the features of signature-based and rich CS descriptors. Basically, there are three strategies to select sample points: (1) Randomly, (2) interesting points and (3) same interval. For the first

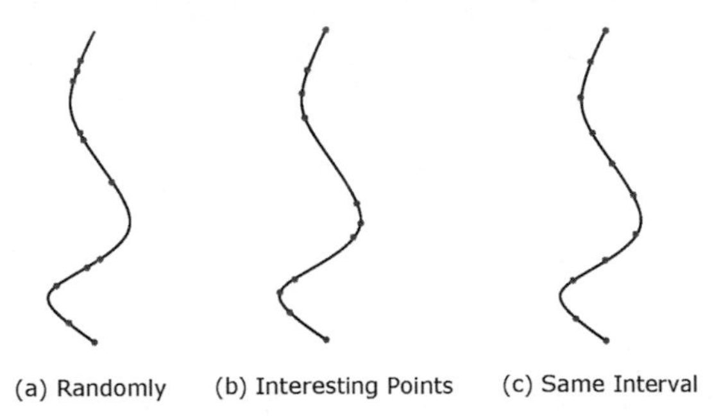

Figure 2.6: Three strategies for selecting sample points (red points) from a CS.

one, the sample points are randomly selected from the sequence of CS points (Figure 2.6 (a)). This strategy is intuitive and with low computational complexity. However, this one is not robust since there is a high probability that some important CS features could not be preserved. Particularly, this strategy could increase the number of mismatches during the point matching process. The main reason is that the location of some sample points are close to each other and it is hard to distinguish them. For the second one, the sample points are selected from the most important parts within a CS (Figure 2.6 (b)). However, comparing to the first one, this strategy has much higher computational complexity [Yan+15a]. In addition, this strategy is also not robust since some meaningful CS points could be ignored if the interesting point selection algorithm cannot be applied properly. For the third one, the sample points are uniformly selected with the same interval. In other words, the number of CS points between two neighbouring sample points are the same (Figure 2.6 (c)). For CS representation, the third strategy is normally employed and proposed for two reasons: (1) This strategy can ensure the equiprobability that most meaningful CS points are preserved. (2) This strategy has low computational complexity. For the simple CS descriptors, the full CS points are normally employed since those simple features can be easily and fast calculated by considering the global distribution of CS points.

2.3.1.1 Simple CS Descriptors

A simple CS descriptor is a scalar that represents a global feature of a CS. CS descriptors in this group are normally generated by considering the global CS geometry. The motivation of simple CS descriptors is that some practical matching problems [PI97] only need simple and coarse-grained features for fast calculations. Thus, it is desired to find descriptors that are both simple and generally applicable. Moreover, a proper combination of descriptors normally offers performance improvements of individuals. In this part, nine simple CS

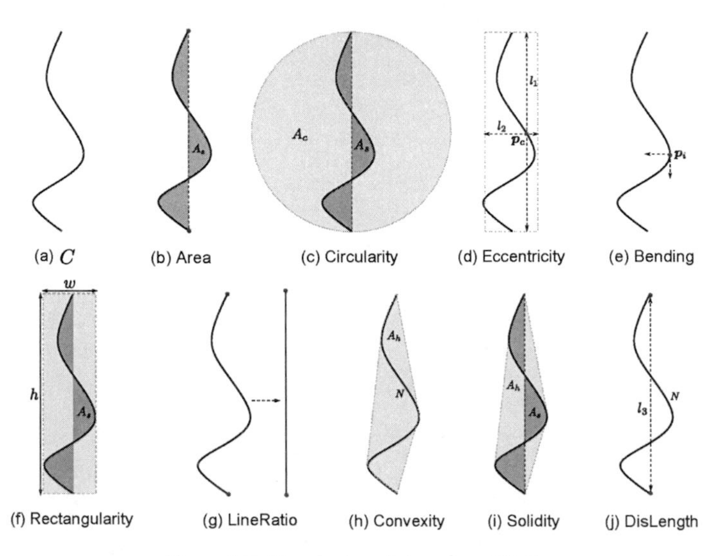

Figure 2.7: Simple descriptors for a CS C.

descriptors from shape survey papers [YIJ08; ZL04] and applications [PI97; YWB74] are surveyed and revised. In general, these descriptors usually can only discriminate CSs with large differences. Therefore, they are not used as standalone descriptors but usually used as filters to eliminate false hits or combined with other rich descriptors.

Area: As shown in Figure 2.7 (b), the area descriptor $\mathbf{f}_1$ [YIJ08; PI97] is calculated as the area A_s (dark grey area) between the straight line (red dotted line) connecting the CS endpoints (red points) and the CS itself (Figure 2.7 (a)). In order to ensure the scale invariance, $\mathbf{f}_1$ is normalised by the length of CS N, that is $\mathbf{f}_1 = A_s/N$.

Circularity: Circularity $\mathbf{f}_2$ [YIJ08; PI97] illustrates how similar the CS C is to a circle. As shown in Figure 2.7 (c), a Circularity $\mathbf{f}_2$ is calculated as A_s/A_c where A_c denotes the area of the minimum CS surrounding circle (light grey area).

Eccentricity: Eccentricity $\mathbf{f}_3$ [YWB74; PI97] can be uniquely defined as the ratio of length of major axis to minor axis that cross each other orthogonally in the middle of the CS. Specifically, the middle point $\mathbf{p}_c$ of CS C (the red point in Figure 2.7 (d)) is firstly located, then Eccentricity is calculated as $\mathbf{f}_3 = l_1/l_2$ where l_1 and l_2 are the lengths of major axis and minor axis to the CS minimum bounding rectangle on $\mathbf{p}_c$, respectively.

Bending: Bending $\mathbf{f}_4$ [ZL04; YWB74] is defined by the average bending energy. It captures the degree of a CS bending energy. For instance, the circle is the shape with the minimum bending energy. Bending is calculated as $\mathbf{f}_4 = \frac{1}{N}\sum_{i=1}^{N} K(i)^2$ where $K(i)$ denotes the curvature of point $\mathbf{p}_i$ (Figure 2.7 (e)) which is approximated by:

$$K(i) = \frac{2|(x_i - x_{i-1})(y_{i+1} - y_{i-1}) - (x_{i+1} - x_{i-1})(y_i - y_{i-1})|}{\sqrt{((x_i - x_{i-1})^2 + (y_i - y_{i-1})^2)((x_{i+1} - x_{i-1})^2 + (y_{i+1} - y_{i-1})^2)((x_{i+1} - x_i)^2 + (y_{i+1} - y_i)^2)}} \quad . \tag{2.1}$$

where $\mathbf{p}_{i-1}$, $\mathbf{p}_i$ and $\mathbf{p}_{i+1}$ are three successive points. $\mathbf{p}_{i-1} = [x_{i-1}, y_{i-1}]$, $\mathbf{p}_i = [x_i, y_i]$ and $\mathbf{p}_{i+1} = [x_{i+1}, y_{i+1}]$.

Rectangularity: Rectangularity $\mathbf{f}_5$ [PI97] presents how rectangular a CS is , i.e. how much the CS fills its minimum bounding rectangle (Figure 2.7 (f)). Rectangularity is calculated as $\mathbf{f}_5 = A_s/(w \cdot h)$ where w and h are the width and height of the CS minimum bounding rectangle.

LineRatio: LineRatio $\mathbf{f}_6$ [YWB74; PI97] uses a straight line as a template, and illustrates how similar a CS is to a straight line (Figure 2.7 (g)). LineRatio is calculated as $\mathbf{f}_6 = h/N$ where h is the height of the CS minimum bounding rectangle.

Convexity: Convexity $\mathbf{f}_7$ [PI97] is defined as the ratio of the convex hull [And79] over that of the CS length. Convexity captures the minimal convex covering of a CS. A straightforward measure for Convexity can be calculated as $\mathbf{f}_7 = A_h/N$ where A_h denotes the CS convex hull area (Figure 2.7 (h)).

Solidity: As shown in Figure 2.7 (i), Solidity $\mathbf{f}_8$ [PI97] describes the extent to which the CS is convex or concave and is defined as A_s/A_h. Solidity is an indicator that captures the concave-convex condition of a CS.

Dislength: Dislength $\mathbf{f}_9$ [Yan+14b] illustrates the skewness power of a CS (Figure 2.7 (j)). It is defined by the ratio between distance of endpoints l_3 and the CS length N.

2.3.1.2 Signature-based CS Descriptors

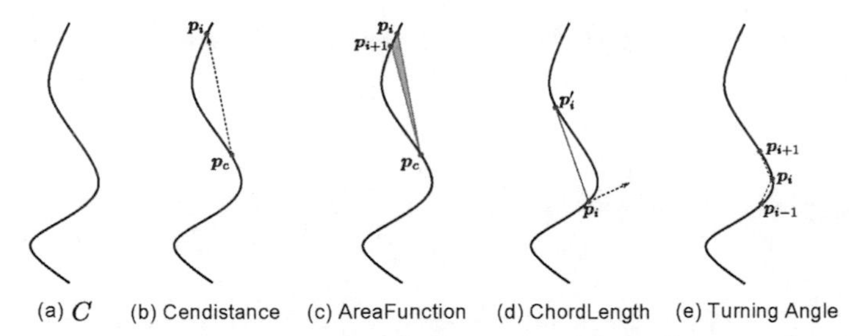

Figure 2.8: Some signature-based descriptors for a CS C.

CS signature is the modified version of Shape Signature that represents a shape by a vector representing various spatial relations among shape boundary points [ZL04]. Signature-based descriptors can capture the perceptual features of CSs and are often combined with some other feature extraction algorithms like Fourier descriptors [VO91; CB84; Arb+90; KLS89; KSP95] and Wavelet Descriptors [Ohm+00; TB97; YLL98]. Different from the simple CS descriptors, the motivation of signature-based CS descriptors is to represent a CS by a higher order of feature vector with low computational cost. Thus, signature-based CS descriptors could offer a proper way for balancing the matching accuracy and speed.

Comcoor: The Comcoor (Complex Coordinates) descriptor $\mathbf{f}_{10}$ [YIJ08] is mainly designed for transforming a CS represented by a two-dimensional sequence (sequence consisting of two-dimensional points) into a one-dimensional sequence. Let $\mathbf{p}_c$ be the middle point for a CS $\mathbf{p}_1, \mathbf{p}_2, \cdots, \mathbf{p}_N$. The Complex Coordinates function is:

$$f_{10,i} = [x_i - x_c] + [y_i - y_c] \quad . \tag{2.2}$$

where $[x_c, y_c]$ is the coordinate of $\mathbf{p}_c$ and $[x_i, y_i]$ is the coordinate of any point in C, $i = 1, 2, \cdots, N$. Finally, $\mathbf{f}_{10} = (f_{10,1}, f_{10,2}, \cdots, f_{10,N})^{\mathrm{T}}$.

Cendistance: The Cendistance (Centroid Distance Function) descriptor $\mathbf{f}_{11}$ [ZL04] is the representation of centroid-based time series. The main characteristics are simplicity and flexibility since the accuracy can be controlled by the density of sample points. Similar to Comcoor, as shown in Figure 2.8 (b), $\mathbf{f}_{11}$ is defined as $(f_{11,1}, \cdots, f_{11,i}, \cdots, f_{11,N})^{\mathrm{T}}$ where $f_{11,i}$ is the angle between $\mathbf{p}_c$ and the CS point $\mathbf{p}_i$.

Tangent: The Tangent (Tangent Angle Function) descriptor $\mathbf{f}_{12}$ [ZL04] is defined as follows:

$$f_{12,i} = \tan^{-1}\left(\frac{y_i - y_{i-w}}{x_i - x_{i-w}}\right) \quad . \tag{2.3}$$

where $\mathbf{p}_i = [x_i, y_i]$ is the ith CS point, and w is a window for controlling the accuracy. Based on preliminary experiments [ZL04], $w = 5$. In such a case, $(i - w)$ is set to 1 if $(1 - w)$ is equal or smaller than 0. Finally, $\mathbf{f}_{12}$ is formed by $\mathbf{f}_{12} = (f_{12,1}, f_{12,2}, \cdots, f_{12,N})^{\mathrm{T}}$. Since the Tangent CS descriptor requires a window for feature generation, it is sensitive to noise. However, it is more flexible than most signature-based CS descriptors since its accuracy can be controlled by changing the size of the window.

Curvature: The Curvature descriptor $\mathbf{f}_{13}$ [ZL04] is very important for capturing salient perceptual characteristics. The Curvature descriptor is calculated as $\mathbf{f}_{13} = (K'(1), K'(2), \cdots, K'(N))^{\mathrm{T}}$ where $K(i)$ be the curvature of a CS point $\mathbf{p}_i$ similar to Bending ($\mathbf{f}_4$). In order to ensure the scale invariance, $K(i)$ is normalised by the mean absolute curvature.

Area Function: The Area Function descriptor $\mathbf{f}_{14}$ [ZL04] is calculated by the triangle area between the middle point $\mathbf{p}_c$ and two consecutive CS points $\mathbf{p}_i$ and $\mathbf{p}_{i+1}$. (Figure 2.8 (c)). In other words, $\mathbf{f}_{14} = (f_{14,1}, \cdots, f_{14,N})^{\mathrm{T}}$ where $f_{14,i}$ is the area of the triangle consisting of $\mathbf{p}_c$, $\mathbf{p}_i$ and $\mathbf{p}_{i+1}$. Area Function is simple but can be used to collect fine-grained features since the distance between $\mathbf{p}_i$ and $\mathbf{p}_{i+1}$ is small and can capture the small deformations of a CS. Moreover, the coarse-grained features are also preserved as the whole areas from $(\mathbf{p}_i, \mathbf{p}_{i+1})$ to a fix point $\mathbf{p}_c$ are collected.

Triangle Area: Different from Area Function, the Triangle Area descriptor $\mathbf{f}_{15}$ [Ala+07] is computed directly from an area of the triangle formed by $\mathbf{p}_{i-t_s}$, $\mathbf{p}_i$ and $\mathbf{p}_{i+t_s}$ where $i \in [1, N]$ and $t_s \in [1, \frac{N}{2} - 1]$. For each CS point $\mathbf{p}_i$, the triangle area is formed by:

$$f_{15,i} = \frac{1}{2} \det \begin{pmatrix} x_{i-t_s} & y_{i-t_s} & 1 \\ x_i & y_i & 1 \\ x_{i+t_s} & y_{i+t_s} & 1 \end{pmatrix} . \tag{2.4}$$

With this equation, when the CS path is traversed in clock-wise direction, positive, negative and zero values of the triangle area mean convex, concave and straight-line points, respectively. Triangle area provides useful information like the convexity/concavity at each CS point. Therefore, this descriptor provides high discrimination capability. Compared to Area Function, Triangle Area is more flexible since the gap between three points can be modified by varying their positions.

Chord Length: The Chord Length descriptor $\mathbf{f}_{16}$ [HS98] is derived by the distance between a CS point and its reference point. As shown in Figure 2.8 (d), the chord length $f_{16,i}$ of $\mathbf{p}_i$ is its shortest distance to the CS point $\mathbf{p}'_i$ so that $\mathbf{p}_i\mathbf{p}'_i$ is perpendicular to the tangent vector at $\mathbf{p}_i$. Finally, $\mathbf{f}_{16} = (f_{16,1}, \cdots, f_{16,N})^{\mathrm{T}}$ where $f_{16,i}$ is normalised by the CS length N for scale invariance. The Chord Length descriptor is robust to fine-grained deformations since chord lengths are calculated using different reference points rather than only one middle point like Area Function, etc. For example, if the middle point of a CS is changed because of deformation or noise, most chord lengths remain the same since each CS point may have a different reference point for calculating its chord length.

Turning Angle: As shown in Figure 2.8 (e), a Turning Angle $\mathbf{f}_{17}$ [DRB10b] represents the angle of $\angle \mathbf{p}_{i-1}\mathbf{p}_i\mathbf{p}_{i+1}$ defined by $\mathbf{p}_i$ and its neighbouring points $\mathbf{p}_{i-1}$ and $\mathbf{p}_{i+1}$. The Turning Angle descriptor $\mathbf{f}_{17}$ can be represented by the whole CS points as a set of feature vectors: $\mathbf{f}_{17} = (f_{17,1}, \cdots, f_{17,N})^{\mathrm{T}}$. Similar to Area Function, Turning Angle captures the fine-grained features of a CS by using adjacent points. However, it has less ability for preserving coarse-grained deformation since the generated turning angles are globally isolated. On the contrary, area functions are not isolated since they are all connected by the middle point of a CS.

2.3.1.3 Rich Descriptors

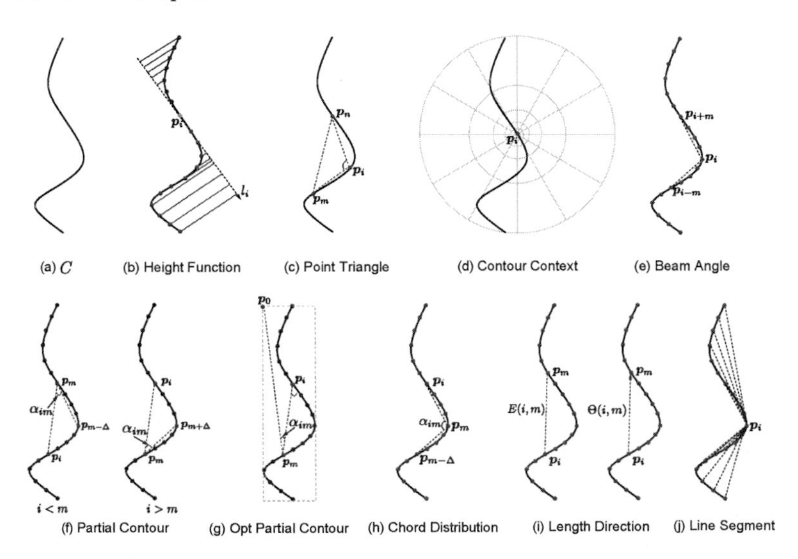

Figure 2.9: Rich descriptors for a CS C.

Rich CS descriptors capture the CS geometrical features in both fine- and coarse-grained levels. Compared to simple and signature-based CS descriptors, the feature vector of a rich descriptor has more dimensions and varieties. Therefore, rich CS descriptors carry more information of the original CS. Specifically, rich descriptors $\mathbf{f}_{18} - \mathbf{f}_{25}$ are usually generated based on feature vector $\mathcal{F}_i$ on each sample point $\mathbf{p}_i$. By collecting $\mathcal{F}_i$ on every sample point $\{\mathcal{F}_1, \mathcal{F}_2, \cdots, \mathcal{F}_N\}$, a rich descriptor can be generated for representing a CS C. In order to simplify the description, the generation of $\mathcal{F}_i$ for each of $\mathbf{f}_{18} - \mathbf{f}_{25}$ is introduced. In contrast, rich descriptor $\mathbf{f}_{26}$ has its own feature structures, it is introduced independently.

Height Function: As shown in Figure 2.9 (b), a height function $\mathcal{F}_i$ [Wan+12b] is defined based on the distances of the other sample points to its tangent line, $i = 1, \cdots, N$. The motivation of this descriptor ($\mathbf{f}_{18}$) is to represent CS points by considering their relations to all other sample points in the same direction.

Point Triangle: The Point Triangle descriptor $\mathbf{f}_{19}$ [Lu+09b] is inspired by Carlsson [TC04] in which they only consider qualitative orientations of each triangle: oriented clock or counterclockwise. The Point Triangle descriptor provides a full quantitative description of each triangle. As shown in Figure 2.9 (c), a Point Triangle descriptor $\mathcal{F}_i$ is a histogram of all triangles spanned by $\mathbf{p}_i$ and all pairs of points $\mathbf{p}_m, \mathbf{p}_n$. Especially, for each triangle, $\mathcal{F}_i$ contains three values: angle $\angle \mathbf{p}_m\mathbf{p}_i\mathbf{p}_n$, distance $\mathbf{p}_i\mathbf{p}_m$ and distance $\mathbf{p}_i\mathbf{p}_n$. This leads to a significant increase in descriptive power since both orientation and distance features are captured in the Point Triangle.

Contour Context: For each point $\mathbf{p}_i$, the Contour Context descriptor $\mathbf{f}_{20}$ [Zhu+08] considers the $N-1$ vectors obtained by connecting $\mathbf{p}_i$ to all other points. The key motivation is that the distribution over relative positions of each CS point is a robust, compact, and highly discriminative descriptor. Similar to Shape Context (SC) [BMP02], in order to capture the geometrical information of $\mathbf{p}_i$, a log-polar histogram is defined by five sections on the radius and 12 sections on the angle (Figure 2.9 (d)). Thereafter, a contour context $\mathcal{F}_i$ for $\mathbf{p}_i$ can be represented as a 60-dimensional feature vector.

Beam Angle: The basic idea of the Beam Angle descriptor $\mathbf{f}_{21}$ [PT10] is to represent each CS point $\mathbf{p}_i$ by a weighted Beam Angle Histogram (BAH). This descriptor represents a CS point using multiple angles with different weights. With this, the Beam Angle descriptor can mitigate the uncertainty in CS representation since it down-weighs the interaction of distant CS points. Specifically, at CS point $\mathbf{p}_i$, the beam angle $\mathcal{F}_i$ is subtended by lines $(\mathbf{p}_{i+m}, \mathbf{p}_i)$ and $(\mathbf{p}_i, \mathbf{p}_{i-m})$ (Figure 2.9 (e)), where $m = 1, \cdots, N'$ and N' is determined experimentally. Therefore, $\mathbf{p}_i$ is represented by N' weighted angles.

Partial Contour: The Partial Contour descriptor $\mathbf{f}_{22}$ [RDB10b] is calculated by the relative orientations between lines that connect the CS sampled points. As shown in Figure 2.9 (f), for a CS point $\mathbf{p}_i$, $\mathcal{F}_i$ is the angle α_{im} which is formed by a line connecting $\mathbf{p}_i$ and $\mathbf{p}_m$, and a line to a third point relative to the position of the previous two points. This third point is chosen depending on the position of the other two points to ensure that the selected point is always inside the CS. This allows them to formulate the descriptor as a self-containing descriptor of any of its parts. The Partial Contour descriptor has several important properties, such as rotation and translation invariance, covering both local and global characteristics, and partial matching based on self-containing.

Opt Partial Contour: The Opt Partial Contour descriptor $\mathbf{f}_{23}$ [PKB11] analyses angles defined by lines connecting a CS-dependent reference point and the CS points. Based on this, both fine- and coarse-grained features can be captured since $\mathbf{f}_{23}$ considers relative angles between CS points and the reference point $\mathbf{p}_0$ which is defined by the upper left corner of the CS surrounding box (See Figure 2.9 (g)). For a CS point $\mathbf{p}_i$, $\mathcal{F}_i$ is formed by the angles $\angle \mathbf{p}_i\mathbf{p}_m\mathbf{p}_0$, where $\mathbf{p}_m$ is also a CS point.

Chord Distribution: A chord is a line joining two points of a region boundary, and the distribution of chords' lengths and angles is often used as a shape descriptor [Coo+92]. The Chord Distribution descriptor $\mathbf{f}_{24}$ [DRB10b] uses chords to exploit the available point ordering information for the subsequent order-preserving assignment matching. In comparison, the contour context descriptor $\mathbf{f}_{20}$ loses all the ordering information due to the histogram binning and does not consider the influence of the local neighbourhood on single point matches. As shown in Figure 2.9 (h), for a CS point $\mathbf{p}_i$, a chord distribution $\mathcal{F}_i$ is computed based on the angles α_{im} which describe the relative spatial arrangement of the sampled points. For a single CS, the angles are calculated over all possible point combinations, yielding the descriptor $\mathbf{f}_{24}$.

Length Direction: As shown in Figure 2.9 (i), for a CS point $\mathbf{p}_i$, the Length Direction descriptor $\mathbf{f}_{25}$ [ML11b] consists of $\mathcal{F}_i$ representing both the length (Euclidean distance in the log space) and direction (four quadrant inverse tangent) of the vector from $\mathbf{p}_i$ to other points in C. The length and direction features are independently saved.

Line Segment: The Line Segment descriptor $\mathbf{f}_{26}$ [MSJB15] is generated based on straight-line segment statistics. As shown in Figure 2.9 (j), for each CS point, the descriptor considers a continuous portion of the CS with length equal to a pre-defined percentage of the CS size. Then, the length of the straight-line segment between the reference point and other CS points is computed using the Euclidean distance. For the set of straight-line segments, statistical movements (average and standard deviation) are calculated. By performing this for different lengths of contour portions, a CS C is represented by a feature vector which describes the CS at different sizes. There are several characteristics of this descriptor: (1) It is simple and intuitive as it is directly generated based on connection lines between CS points. (2) It preserves the coarse-grained features hierarchically using different scales. (3) Built on the feature vectors, the distance between CSs can be quickly calculated by vector distance.

2.3.2 Open Curve Matching using Contour Segments

In the first part, the methods for calculating the distance $\mathcal{D}(C_1, C_2)$ between two CSs C_1 and C_2 are introduced, depending on types of CS descriptors presented in the previous section. After that, the open curve matching approach is introduced in the second part. In order to ensure the scale invariance (see Figure 2.12), all feature values are normalised before the matching process.

2.3.2.1 Contour Segment Matching

Simple Descriptors: Since the simple descriptors $\mathbf{f}_1, \mathbf{f}_2, \cdots, \mathbf{f}_9$ are scalar, their feature values can be represented by $f_1, f_2, \cdots, f_9$, respectively. The distance $\mathcal{D}(C_1, C_2)$ is calculated by:

$$\mathcal{D}(C_1, C_2) = \frac{|f_{1,j} - f_{2,j}|}{f_{1,j} + f_{2,j}} \quad . \tag{2.5}$$

where $f_{1,j}$ and $f_{2,j}$ are the feature values of the jth simple descriptor between C_1 and C_2, respectively. Note that simple descriptor values significantly vary depending on the CSs.

Thus, it is needed to make their differences independent of their descriptor values. To this end, Equation (2.5) is designed to normalise the difference between two simple descriptor values. Since all feature values of simple descriptors are positive and normalised, Equation (2.5) is defined as a distance method.

Signature-based Descriptors: There are two methods for calculating $\mathcal{D}(C_1, C_2)$ between C_1 and C_2 represented by signature-based descriptors: Point matching and vector distance. Let $\mathbf{f}_{1,j} = (f_{1,j,1}, \cdots, f_{1,j,N})^{\mathrm{T}}$ and $\mathbf{f}_{2,j} = (f_{2,j,1}, \cdots, f_{2,j,N})^{\mathrm{T}}$ be the jth signature-based CS descriptors for C_1 and C_2, respectively.

With the point matching method, the differences $d(f_{1,j,m}, f_{2,j,n})$ between $f_{1,j,m}$ in $\mathbf{f}_{1,j}$ and $f_{2,j,n}$ in $\mathbf{f}_{2,j}$ $(m, n = 1, 2, \cdots, N)$ are firstly calculated. Then, a matrix of differences between all CS points in C_1 and C_2 is generated. In order to find an optimum match of CS points between C_1 and C_2, the matching algorithms like DTW [SC04], DP [Ric54] and Hungarian [Kuh55] are used on the matrix. Based on these algorithms, CS point correspondences and matching costs can be obtained as the matching results. Specifically, in order to avoid the brute-force approach [BC94] of the standard DTW algorithm, the FastDTW [SC04] is employed for CS point matching. For the DP, the solution proposed by Sellers [Sel80] is employed to reduce the time complexity of the traditional approach [Ric54]. The Hungarian algorithm [Kuh55] solves the assignment problem in a weighted bipartite graph. With this approach, the correspondence between CS points is generated by minimising the global cost between CS point distances. The resulting distance values of the matched CS points can be denoted as $s_1, \cdots, s_N$ and the similarity between C_1 and C_2 is calculated as the mean value.

With vector distance computation, the distance between $\mathbf{f}_{1,j}$ and $\mathbf{f}_{2,j}$ is directly calculated without considering any point matching. In particular, the following distance measures are employed: (1) Correlation [YK8], (2) HI [RTG00], (3) χ^2-Statistics [RTG00] and (4) Hellinger (or Bhattacharyya Coefficient) [Bha46]. These distance measures are selected based on their simplicity and applicability evaluated in [Kar+15]. Since there are many other existing distance measures, in Section 8.2, their adoptions are discussed as future work.

Rich Descriptors: Since feature vectors of rich descriptors $\mathbf{f}_{18}, \cdots, \mathbf{f}_{26}$ are not uniformed, they have their own way of matching CS points as well as calculating CS distances: (1) For the Height Function descriptor $\mathbf{f}_{18}$, to match two CSs C_1 and C_2, the distances between any pair of points are computed by the weighted difference of their height features [Wan+12b]. Then, a cost matrix is generated. Here, high weights are put on CS points near to the center to tolerate CS deformations. Finally, a matching algorithm like Hungarian, DP or DTW is applied on the cost matrix to get the similarity between C_1 and C_2. (2) For the descriptors $\mathbf{f}_{19}$ - $\mathbf{f}_{24}$, they are all constructed by the histograms of CS points. Thus, a cost matrix is obtained by computing the distance between any pair of points using the histogram intersection. Then, the overall similarity between C_1 and C_2 is calculated by applying a matching algorithm like Hungarian, DP or DTW to the cost matrix. (3) For the Length Direction descriptor $\mathbf{f}_{25}$, the distance between C_1 and C_2 is calculated by the method in [ML11b] in which the distance and orientation matrices are fused to get an affinity matrix that represents the similarities of corresponding point pairs. Then, the optimal correspondence and overall similarity between C_1 and C_2 are calculated with Hungarian, DP or DTW. (4) For Line Segment $\mathbf{f}_{26}$, this descriptor is organised into vectors, so the vector distance methods can be used directly on it.

2.3.2.2 Open Curve Matching

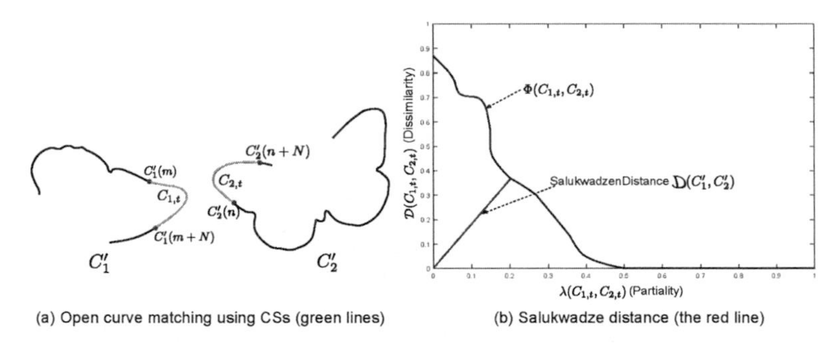

(a) Open curve matching using CSs (green lines) (b) Salukwadze distance (the red line)

Figure 2.10: Open curve matching and the open curve similarity method.

Given two open curves C'_1 and C'_2 with sizes N_1 and N_2, respectively, searching similar parts between C'_1 and C'_2 is equivalent to finding two CSs C_1 and C_2 with size N starting at the curve points $C'_1(m)$ and $C'_2(n)$ which yield the smallest distance $\mathcal{D}(C_1, C_2)$, $m \leq N_1$, $n \leq N_2$, $N \leq \min(N_1, N_2)$ (Figure 2.10 (a)). Therefore, the problem is reduced to searching the optimal (m, n, N) combination which minimises the CS distance [DRB10b]. Note that the similarity between C_1 in C'_1 and C_2 in C'_2 depends on their length N. In other words, a smaller N often leads to a higher similarity. To avoid this undesirable effect of N, the Pareto-framework [Bro+09; DRB10b] is employed for quantitative interpretation of partial similarity. The Pareto-framework illustrates a way to select a multi-constrained path that can meet the optimal requirements.

As shown in Figure 2.10 (b), two quantities are defined: partiality $\lambda(C_{1,t}, C_{2,t})$, which describes the length of the CS (the higher the value, the smaller the CS), and the distance $\mathcal{D}(C_{1,t}, C_{2,t})$ which measures the dissimilarity, $t = 1, 2, \cdots, H$, with H being the number of similar parts between C'_1 and C'_2. A Pareto optimum is defined by $\Phi(C_{1,t}, C_{2,t})$ which is a pair of partiality and dissimilarity values that fulfill the criterion of the lowest dissimilarity for the given partiality. Finally, to compute a similarity value between C_1' and C_2', the so-called Salukwadze distance is employed which measures the minimum distance from the origin (0,0) to the point on the Pareto optimum. The Salukwadze distance is then returned as the open curve similarity value. This similarity value will be used to evaluate CS descriptors in an open curve retrieval scenario.

2.3.3 Evaluating Contour Segment Descriptors

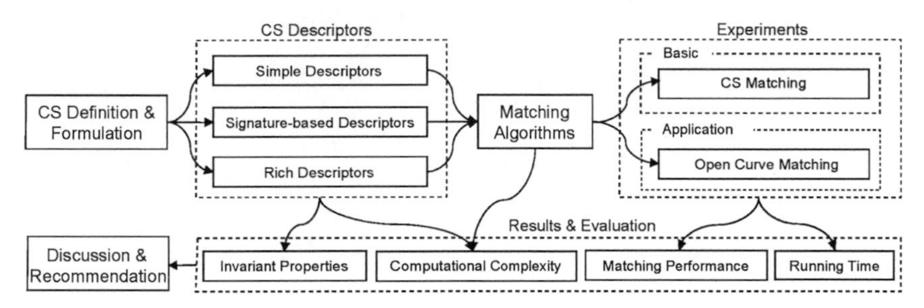

Figure 2.11: Pipeline for evaluating CS descriptors.

Based on the introduction of CS descriptors and the matching algorithms, the performance of the aforementioned descriptors are evaluated in this part. Figure 2.11 illustrates the pipeline of the evaluating steps. Specifically, the evaluation is applied by taking four properties into account. Firstly, a taxonomy of invariance properties are analysed and compared. Secondly, the computational complexity of both feature generation and CS matching on 26 CS descriptors are theoretically analysed. Thirdly, the matching performance of CS descriptors is analysed using different combinations of CS descriptors and matching algorithms via two experiments. Lastly, the runtime in the CS matching experiment is evaluated and compared. With discussions and observations on the four properties above, the recommendation is addressed for different application scenarios by balancing the matching accuracy and speed.

2.3.3.1 Evaluation Design

Accurate matching requires effective CS descriptors to find perceptually similar curves from a database, irrespective of their rotations, translations and scales. A desirable CS descriptor should have a stable performance on different types of datasets. Low computation

Table 2.1: Experiment settings for evaluating the matching performance of CS descriptors. L1, L2, L3 and L4 denote the sampling densities that are used for generating CS descriptors. NULL means this part is not considered in the evaluations. For each type of descriptors, the evaluation is applied in two scenarios: CS matching and open curve matching.

<table>
<tr><th>Descriptors</th><th colspan="2">Distance Functions</th><th>Matching Algorithms</th><th>Lengths</th><th>Scenarios</th></tr>
<tr><td>Simple
($\mathbf{f}_1$ - $\mathbf{f}_9$)</td><td colspan="2">Difference Value</td><td>NULL</td><td>L1, L2
L3, L4</td><td>(1) CS
(2) Open Curve</td></tr>
<tr><td rowspan="2">Signature
($\mathbf{f}_{10}$ - $\mathbf{f}_{17}$)</td><td colspan="2">Difference Value</td><td>(1) DTW (2) DP
(3) Hungarian</td><td rowspan="2">L1, L2
L3, L4</td><td rowspan="2">(1) CS
(2) Open Curve</td></tr>
<tr><td colspan="2">(1) Correlation (2) HI
(3) χ^2-Statistics (4) Hellinger</td><td>NULL</td></tr>
<tr><td rowspan="2">Rich
($\mathbf{f}_{18}$ - $\mathbf{f}_{26}$)</td><td>$\mathbf{f}_{18}$ - $\mathbf{f}_{25}$</td><td>Proposed Distances</td><td>(1) DTW (2) DP
(3) Hungarian</td><td rowspan="2">L1, L2
L3, L4</td><td rowspan="2">(1) CS
(2) Open Curve</td></tr>
<tr><td>$\mathbf{f}_{26}$</td><td>(1) Correlation (2) HI
(3) χ^2-Statistics (4) Hellinger</td><td>NULL</td></tr>
</table>

complexity is also an important factor of a CS descriptor. Considering these requirements, three types of experiments are established.

Experiment 1: Invariance Properties: Descriptors are often classified according to their invariance levels to certain geometrical transformations. Here, three invariance properties of each CS descriptor are mainly assessed: rotation, scaling and translation invariances. As shown in Figure 2.11, rotation and scaling invariances mean that the CS features remain the same even when the CS is rotated and rescaled. Translation invariance means that two CSs can still be correctly matched even when every CS point is moved to a constant distance in a specified direction.

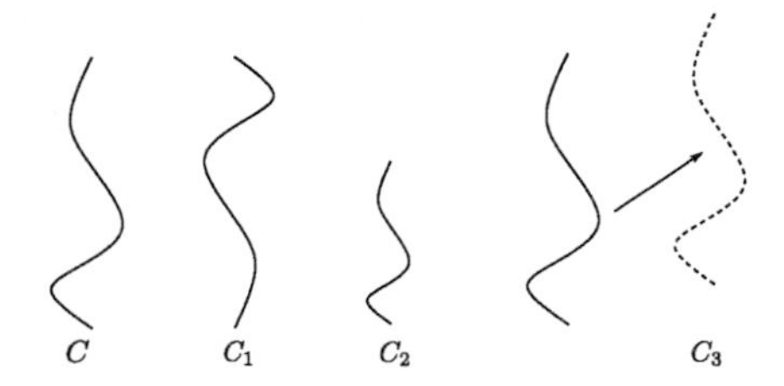

Figure 2.12: Invariance properties of a CS C: rotation (C_1), scaling (C_2) and translation (C_3).

Experiment 2: Matching Performance: As illustrated in Table 2.1, CS descriptors are evaluated under multiple settings. For all descriptors, their features are generated using four sampling densities for selecting sample points from a CS: L1 = 25%, L2 = 50%, L3 = 75%

and L4 = 100%. This strategy aims to check the influence of sample point density to the CS matching performance. With different combinations of distance functions, matching algorithms and CS lengths (sampling densities), the experiments are conducted on two scenarios: CS matching and open curve matching.

Experiment 3: Computational Complexity: The computational complexity of each CS descriptor is examined in terms of the time complexity analysis and the real runtime. The computational time complexities are theoretically analysed for the generation and matching of each CS descriptor [HS11]. The real runtime is evaluated based on the full retrieval time in a specific dataset.

2.3.3.2 Recommendation

Table 2.2: Recommended CS descriptors for different open curve matching requirements.

Best Accuracy	Promising Accuracy and Less Runtime	Fast Speed and Relatively Promising Accuracy
Point Triangle ($\mathbf{f}_{19}$) + DP	Partial Contour ($\mathbf{f}_{22}$) + DP	Comcoor ($\mathbf{f}_{10}$) + Hellinger
	Chord Distribution ($\mathbf{f}_{24}$) + DP	Cendistance ($\mathbf{f}_{11}$) + Hellinger

Based on the experiments and observations in Section 6.1.1, the following recommendations are drawn (Table 2.2): (i) When choosing a CS descriptor for open curve matching with time complexity not being of primary importance, the best choice is Point Triangle ($\mathbf{f}_{19}$) with DP [Ric54] since it is scale, rotation and translation invariant (Table 6.1) and robust to CS length changing (Table 6.5 and Table 6.10). Moreover, it achieves promising results in both CS and open curve matching scenarios (Table 6.13). (ii) Partial Contour ($\mathbf{f}_{22}$) and Chord Distribution ($\mathbf{f}_{24}$) with DP are the best choices to obtain a stable and promising performance while taking less computational time, as shown in Table 6.13. (iii) If a fast open curve matching and a relatively promising result are required, Comcoor ($\mathbf{f}_{10}$) and Cendistance ($\mathbf{f}_{11}$) with the Hellinger [Bha46] vector distance method are the best choice, for which the fast runtime for feature generation and matching can be ensured while the matching performance in both scenarios is not the best but still promising. The detailed performance changes by varying the combinations between CS and matching methods are illustrated in Table 6.2 ~Table 6.11.

2.4 Shape Contour Detection by Open Curve Matching

Based on the proposed open curve matching method, the object shape contour detection and generation methods are introduced in this part. As shown in Figure 2.2, an object shape contour is detected by partially matching open curves to a single shape prototype (model) of the category under the constraints of length and dissimilarity. After that, the eligible open curves are aggregated to form the shape contour of an object. The works introduced in this part are inspired by [RDB10a]. Different from their original works, as recommended in Table 2.2, the Point Triangle ($\mathbf{f}_{19}$) CS descriptor with DP is used for open curve matching since it has a stable and high performance during the evaluation.

2.4.0.3 Open Curve Selection

Based on an open curve book and a shape contour model, the main purpose of this part is to find proper open curves which are similar to the model contour. Since the detected shape and the model shape may have different geometry, their shape contours may only be partially similar to each other. Moreover, the open curves that form the detected shape may be connected to some irrelevant curves. Thus, for any two open curves from the open curve book and the model contour, the aim of matching is to identify parts which are similar to each other (see Figure 2.13).

Figure 2.13: Searching similar parts among four open curves using CS (marked with the red colour) matching. Some open curves may have multiple similar parts, but here only one is marked for better visualisation.

Based on the open curve matching method introduced in Section 2.3.2.2, given two open curves, their sub-blocks are compared using CS to find all matching possibilities and lengths. Thus, the problem is reduced to searching the optimal (m,n,N) triplet which minimises the CS dissimilarity. m and n identify the starting points of the match in the reference and the model open curves, respectively. N defines the length of the match. All matching triplets (m,n,N) are stored in a correspondence tensor $\Gamma_{(m,n,N)}$ with a three-dimensional (3D) dissimilarity format. Note that this tensor fully defines all possible correspondences between the reference and the model curves. Since there may be many overlapping and repetitive matches, the goal is to find the longest and most similar parts and merge repetitive matches instead of retrieving all individual matches. In order to do so, as introduced in [RDB10a], some properties of the 3D dissimilarity and correspondence tensor $\Gamma_{(m,n,N)}$ are firstly outlined:

I An element (m,n,N) has a dissimilarity value based on CS matching.

II Length variations (m,n,N') with $N' < N$ define the same correspondence, yet shorter in length.

III Diagonal shifts in the indices $(m+1,n+1,N)$ also represent the same match, yet one starting point later.

IV Unequal shifts $(m+1,n,N)$ define a different correspondence, however very similar and close.

V Due to occlusions or noise, multiple matches per edge contour may exist (see Figure 2.13).

VI Matches near to the end of each open curve have a maximal length given by the remaining points in each curve sequence.

Based on these properties, a matching criterion [RDB10a] is defined to deliver the longest and most similar matches. Specifically, (a) finding valid correspondences satisfying the constraints on length and dissimilarity, (b) merging all valid correspondences to obtain the longest combination of the included matches (property II) and (c) selecting the minimal dissimilarity of matches in close proximity (property IV). The details are as follows:

Firstly, a function $\mathcal{L}(m,n,N)$ which gives the lengths at any valid correspondence tuple (m,n) as:

$$\mathcal{L}(m,n,N) = \begin{cases} N & \text{if } \Gamma_{(m,n,N)} \leqslant \mathcal{D}_{lim} \text{ and } N \geqslant N_{lim} \\ 0 & \text{otherwise} \end{cases} \quad . \tag{2.6}$$

where the value at $\mathcal{L}(m,n,N)$ is the length of a valid matched part which has a dissimilarity score below the limit $\mathcal{D}_{lim}$ and a minimal length limit of N_{lim}. This function is used to define a subset of longest candidates $\Psi_{m,n}$ by

$$\forall_{m,n} : \underset{N \in N_1, N_2}{\arg\max} \mathcal{L}(m,n,N) \quad . \tag{2.7}$$

where N_1, N_2 are the lengths of compared open curves. $\Psi_{m,n}$ is a subset of $\Gamma_{(m,n,N)}$ containing the longest matches at each correspondence tuple (m,n). This set contains matches for every possible correspondence given by the constraints on dissimilarity and matching positions (see property II, VI). However, it is worth reducing this to only the local maxima (conserving property IV). Since the set $\Psi_{m,n}$ can now be considered as a 2D function, the composed curve c which satisfied $\Psi_{m,n} > 0$ can be found. The final set of candidates $\Upsilon_{(m,n,N)}$ are the maxima per connected component and is defined as

$$\forall c_i \in C' : \underset{\Gamma_{(m,n,N)}}{\arg\max} (\Psi_{m,n} \in c_i) \quad . \tag{2.8}$$

where $\Upsilon_{(m,n,N)}$ holds the longest possible and most similar matches given the constraints on minimum dissimilarity $\mathcal{D}_{lim}$ and minimal length N_{lim}. Here $c_i \in C'$ is added to ensure a composed curve c_i still fulfills the constraints of open curve C' which is established in Section 2.2.

Figure 2.14: Shorter matches (the independent red and green parts) are merged to longer matches (the combined red and green part) by analysing the whole tensor $\Gamma_{(m,n,N)}$.

It is noteworthy that shorter matches in some parts are possible, but are directly merged to longer and more discriminative matches by analysing the whole tensor (e.g. Figure 2.14). Furthermore, the obtained matches are local maxima concerning dissimilarity scores. This provides an efficient open curve set leading to discriminative matches and reduced runtime.

2.4.0.4 Hypothesis Selection

Built on matching in the previous part, a set of matched and merged open curves are selected as the candidate curves, which have to be combined to form object localisation and contour detection hypotheses (see Figure 2.2). The following parts describe how matched open curves are grouped for object contour location and detection hypotheses.

Open Curve Aggregation: So far, a set $\Upsilon_{(m,n,N)}$ is collected which contains the matched parts of open curves detected in a query image which are highly similar to the provided model contour. Each match has a certain dissimilarity and length. Further, as introduced in [RDB10a], each matched open curve can be mapped to its model contour and estimate the object centroid from the given correspondence tuple (m,n,N). Specifically, the matched open curves are clustered by analysing their corresponding center votes and their scale by mean-shift mode detection. Based on the vote and scale analysis, the aggregation is applied to identify groups of open curves which compliment each other and form object location hypotheses.

Hypothesis Ranking: Since $\Upsilon_{(m,n,N)}$ contains matched open curves from different locations in an image, there could be more than one hypothesis appearing after the aggregation. In such a case, the most confident one should be selected as the final shape contour detection result. In order to do so, all obtained hypotheses are ranked according to their confidence. For this purpose, a ranking method proposed in [RDB10a] is employed. Specifically, the ranking score ς_{COV} is calculated based on the amount of the model contour that is covered by the matched open curves. In order to ensure scale invariance, ς_{COV} is normalised by the number of contour points in the model contour. The coverage score ς_{COV} provides a value describing how many parts are matched to the model contour for the current hypothesis. Based on the ranking, the hypothesis with the highest ς_{COV} value is returned as the final contour detection result.

2.5 Summary

This chapter introduced an intuitive and simple shape contour detection algorithm for the purpose of shape based object localisation and shape generation. It is based on the idea of open curve matching between a model contour and an open curve book which is generated from a given image. In order to generate a robust open curve book, Canny edges of a given image are filtered into an open curve map and then an open curve reduction process is applied to remove small and useless curves. This process is addressed to solve the problem of background clutter. In order to stabilise the detection accuracy and reduce the matching time, 26 different CS descriptors are evaluated specially for open curve matching. Based on the evaluation results, the recommendation is addressed for different open curve matching scenarios by balancing the matching accuracy and speed. For shape contour detection, candidate open curves in an open curve book are matched and selected to form different hypotheses followed by a voting scheme. The final detection result is the hypothesis which achieved the highest confidence score during the voting. Based on the detected shape contour, an object shape is generated by properly filling the hollow space. The main scientific contributions of this part include: (1) Two annotated datasets are specially designed and proposed for CS descriptor evaluation. (2) Five combinations of CS descriptors and matching algorithms are proposed for the task of shape generation in three scenarios. In Section 6.1, two types of experiments are applied to report the CS evaluation as well as the shape generation results.

Chapter 3

Shape Representation with Coarse-grained Features

Coarse-grained shape descriptors mainly capture the shape features from global and simple object deformations. The motivation is built on the fact that in some shape-based applications [Yan+14a; Grz+13], only coarse-grained features are required for object matching and classification. Therefore, these features are normally used to distinguish shapes with large differences (see Figure 3.1). Moreover, since coarse-grained features have the property of easy generation and fast matching [ZL04], the correlated descriptors are usually used as filters to eliminate false hits or combined with other descriptors to discriminate shapes [LLE00].

Figure 3.1: Sample shapes with large differences.

However, for shape description, there is always a trade-off between accuracy and efficiency. On the one hand, shape should be described as accurately as possible; on the other hand, a shape descriptor should be as compact as possible to simplify indexing and retrieval. In order to do so, one typical approach is to improve the adaptability of coarse-grained shape descriptors for some specific datasets. Another possible way is to capture more shape features without adding too much computational complexity. Therefore, the proposed shape representation approaches feature a shape by considering both possibilities.

Shape representation methods can be generally classified into two classes: contour-based methods and region-based methods. The classification is based on whether shape features are extracted from the contour only or are extracted from the whole shape region [ZL04]. Specifically, contour-based methods only exploit shape boundary information while region-based approaches take all the pixels within a shape to obtain the coarse-grained shape features. In the following sections, one shape representation method is proposed under each class. In Section 6.2, the correlated experiments and results are reported and discussed.

3.1 Techniques for Coarse-grained Feature Generation

For coarse-grained shape representation, a detailed review of those techniques along with their categorical classification is given in [ZL04; Ama+11]. Common coarse-grained shape descriptors include area, circularity (perimeter^2/area), eccentricity (length of major axis/length of minor axis), major axis orientation, and bending energy [YWB74]. Other descriptors including convexity, ratio of principle axis, circular variance and elliptic variance have been proposed by Peura *et al.* [PI97]. These descriptors usually can only discriminate shapes with big deformations. In order to improve the discriminative ability, Duci *et al.* [Duc+03] suggested embedded closed planar contours that possess a linear signature as a subset of harmonic functions of which the original contour is a zero level-set. B-Splines [CW94; GT00; WT04] are widely used for shape representation. More recently, Chain codes [Yu+10; AS05] have also been used for shape representation though they are not considered reliable for shape matching [AS05]. The main reason is that they suffer from discretisation errors with respect to rotation and scale [JB14]. Nguyen *et al.* [TNOL13] propose a shape-based local binary descriptor for object detection that has been tested in the task of detecting humans from static images.

Compared to the representation methods above, signature-based shape descriptors normally have a higher ability to discriminate coarse-grained deformations. A shape signature represents a shape by a one dimensional function derived from shape boundary points [ZL04]. Here the signature-based descriptors are classified into coarse-grained methods since most of them are generated and saved by considering individual or neighbouring shape boundary points. In such a case, a shape topology cannot be preserved. Many signature-based shape descriptors exist [VO91; ZL02], including centroidal profile, complex coordinates, centroid distance, tangent angle, cumulative angle, curvature and chord-length, etc. These descriptors are usually normalised into being translation and scale invariant. In order to compensate for orientation changes, shift matching is applied during the shape matching process. Besides the high matching cost, shape signatures are sensitive to noise, and slight changes in the boundary can cause large errors in matching [ZL02]. Therefore, it cannot be used directly to describe shapes. In practice, additional processing is necessary to increase its robustness. For example, a signature descriptor can be simplified by quantising the signature into a signature histogram, which is rotationally invariant [ZL04].

Different from the above descriptors, the proposed coarse-grained shape descriptors can be adapted to different datasets by enhancing specific shape features. Moreover, the proposed descriptors are invariant to shape rotation, scaling and transformation. Lastly, benefiting from the structures of their feature vectors, different fast matching algorithms can be directly applied.

3.2 Contour-based Method

In this part a coarse-grained shape descriptor based on shape contour is introduced. With the Discrete Curve Evolution (DCE) approach, a shape contour is partitioned into open curves by partition points (Figure 3.2). Each open curve is described by a rotation and scale invariant 12-dimensional feature vector. Then a shape is represented by the feature vectors of all open curves. There are two differences between the open curves in Section 2.3 and in this part. Firstly, in Section 2.3, open curves are generated from an image. The open curves in this part are generated from a shape contour. Secondly, in Section 2.3, open curves are matched for shape contour detection. In this part, the generated open curves are employed for shape matching. In the following parts, the DCE method for generating contour partition points is firstly introduced. After that, the proposed shape representation method is addressed.

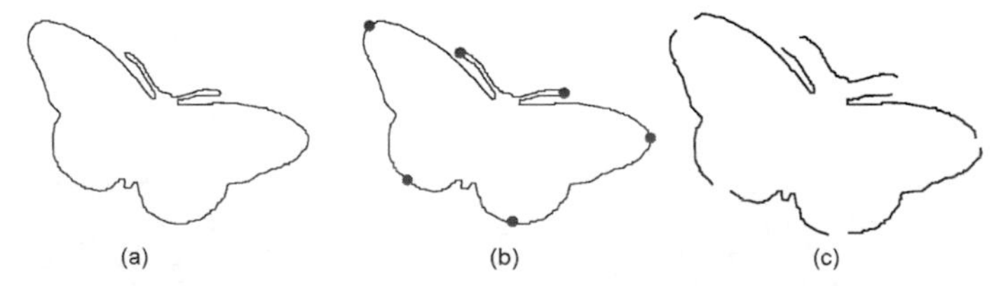

Figure 3.2: General idea of the proposed method. Partition points (red points in (b)) are detected by the DCE method.

Discrete Curve Evolution: The Discrete Curve Evolution (DCE) was introduced by Latecki et al. [LL99] in 1999. It is based on the fact that a shape contour is easily distorted by digitisation noise and segmentation errors. Therefore, it is desirable to eliminate the distortions while at the same time preserving the perceptual appearances sufficient for object recognition. The DCE method accomplishes this goal by simplifying the shape. The core idea of DCE relies on the assumption that the boundary of an object shape D can be represented as a finite polygon P without any loss of information.

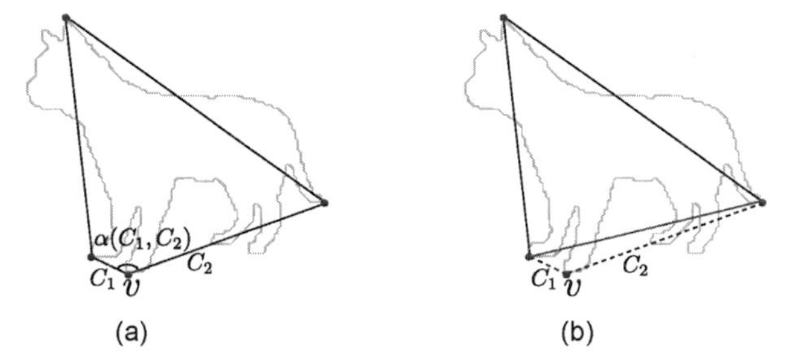

Figure 3.3: Polygon simplification. As vertex v has the smallest contribution with Equation (3.1), its consecutive line segments C_1, C_2 are replaced by a single line segment (red line in (b)).

Below, the detailed implementation of DCE for generating P is introduced. As shown in Figure 3.3, a pair of consecutive line segments C_1, C_2 is replaced by a single line segment that connects the endpoints of $C_1 \cup C_2$. DCE produces a sequence of simpler polygons $P = P_n, P_{n-1}, \cdots, P_3$ so that P_i is obtained by removing a single vertex v from P_{i+1}, $i \in [3, n)$ and $i \in \mathbb{N}$. Here, v is regarded as having the smallest contribution based on the following measure ψ:

$$\psi(C_1, C_2) = \frac{\alpha(C_1, C_2) l(C_1) l(C_2)}{l(C_1) + l(C_2)} \quad . \tag{3.1}$$

where $\alpha(C_1, C_2)$ is the angle of the corner consisting of C_1 and C_2. The length function l is normalised with respect to the total length of lines constituting the polygon. Based on ψ, while the the value of $\psi(C_1, C_2)$ becomes higher, the contribution of $C_1 \cup C_2$ to the polygon is larger. A few stages of polygon simplification are illustrated in Figure 3.4.

Object Shape Representation Using Open Curves: Based on the DCE method described above, the object contour is divided into N open curves by the polygon vertexes. However, during the polygon simplification process, each polygon may have a different number of vertexes. In order to solve this problem, a fixed DCE step can be selected to generate partition points for all shapes. The theoretical description is that the shape simplification power is fixed. This is reasonable since it is challenging to find a general simplification power for covering all shape deformation types. A deeper discussion of this phenomenon is also addressed in Section 4.3.

For each open curve, a 12-dimensional meaningful feature vector $\mathbf{c}_n$ ($n = \{1, \cdots, N\}$) is

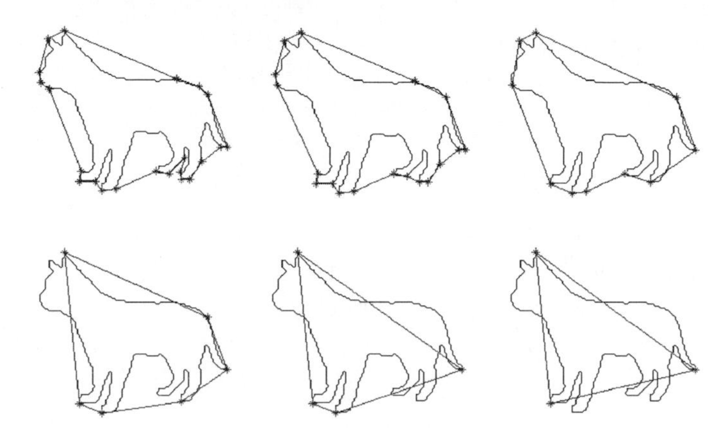

Figure 3.4: Polygon P (in blue line) and vertex v (in red asterisk) changes along with a few stages of DCE.

extracted, whereas its first feature is equal to the number of open curves resulting from the whole object ($f_{n,1} = N$). The Euclidean distance of open curve endpoints and the total number of pixels in an open curve determine $f_{n,2}$ and $f_{n,3}$, respectively. These two features are able to express how much an open curve differs from a straight line. In order to distinguish open curves of the type presented in Figure 3.5 (a) from those of the type depicted in Figure 3.5 (b), the area between the straight line connecting the open curve endpoints and the open curve itself (marked as grey in Figure 3.5) is used as the fourth feature ($f_{n,4}$).

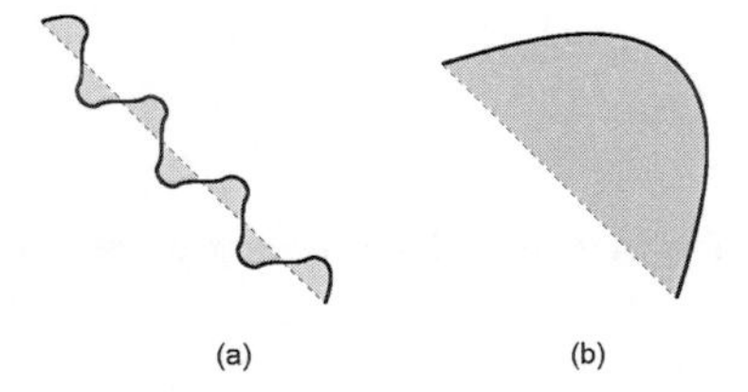

Figure 3.5: Since $f_{n,2}$ and $f_{n,3}$ are equal for open curves (a) and (b), a fourth feature $f_{n,4}$ corresponding to the area depicted in grey is introduced.

Before computing remaining features, each open curve is transformed into a normalised vertical orientation (i.e., so that its endpoints are vertically aligned) to ensure rotation invariance of the object contour representation (see Figure 3.6). From the two possible results of such a normalising transform, the open curve with the majority of points lying on the right side of the straight line connecting its endpoints is selected for further processing. For computing further features $f_{n,5}, f_{n,6}, \ldots, f_{n,12}$, the bounding box of the whole open curve as well as the three equally high sub-boxes are used (Figure 3.6). Here the bounding box is subdivided into 3 equally high sub-boxes, this is based on the trade-off between configuration and fineness of subdivision. If a bounding box is decomposed into more sub-boxes, each sub-segment located in sub-boxes tends to be similar. This could give rise to mis-corresponding during the curve matching process. Based on experiments in [Fei+14; Yan+14b], 3 sub-boxes selection achieves the best performance in terms of accuracy and robustness.

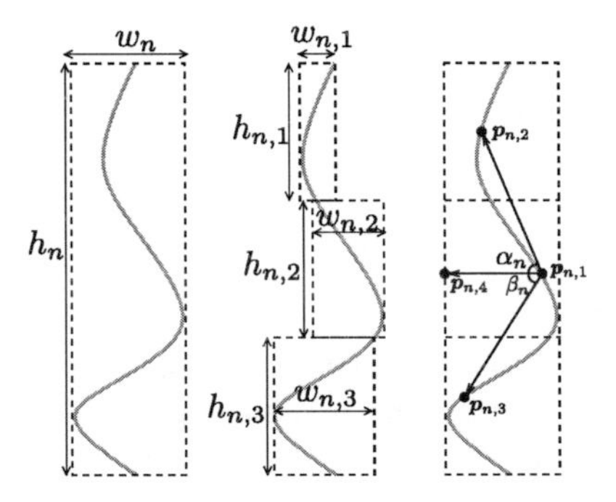

Figure 3.6: Open curve bounding box and equally high sub-boxes ($h_{n,1} = h_{n,2} = h_{n,3}$) used for feature extraction: $\mathbf{p}_{n,1} \rightarrow \mathbf{p}_{n,4}$ is oriented vertically; $\mathbf{p}_{n,2}$, $\mathbf{p}_{n,1}$, and $\mathbf{p}_{n,3}$ are centre pixels of the top, middle, and bottom curve sub-segments, respectively.

$$\begin{array}{llll} f_{n,5} = \frac{h_n}{w_n} & f_{n,6} = \frac{h_{n,1}}{w_{n,1}} & f_{n,7} = \frac{h_{n,2}}{w_{n,2}} & f_{n,8} = \frac{h_{n,3}}{w_{n,3}} \\ f_{n,9} = \frac{w_{n,3}h_{n,3}}{w_{n,1}h_{n,1}} & f_{n,10} = \frac{w_{n,2}h_{n,2}}{w_{n,1}h_{n,1}} & f_{n,11} = \alpha_n & f_{n,12} = \beta_n \end{array} \quad . \tag{3.2}$$

Here $f_{n,2}, f_{n,3}$ and $f_{n,4}$ are divided by a half of the bounding box perimeter for scale invariance. For other elements, scale normalisation is not required since they are ratio or angle values. Finally, an open curve can be represented by $\mathbf{c}_n$:

$$\mathbf{c}_n = (f_{n,1}, f_{n,2}, \cdots, f_{n,12})^{\mathrm{T}} \quad . \tag{3.3}$$

where $n = \{1, \cdots, N\}$. The whole object shape D can be represented as a collection of feature vectors describing its contour:

$$D = \{\mathbf{c}_1, \mathbf{c}_2, \cdots, \mathbf{c}_n \cdots, \mathbf{c}_N\} \quad . \tag{3.4}$$

3.3 Region-based Method

In this part, an intuitive and simple coarse-grained shape descriptor with low computation complexity is proposed. The biggest motivation of this descriptor is that for some shapes, partition points are hard to be generated. For instance, the DCE method cannot be applied on a circle since all contour points could be vertices. Therefore, the proposed shape descriptor in this section is directly calculated based on shape regions.

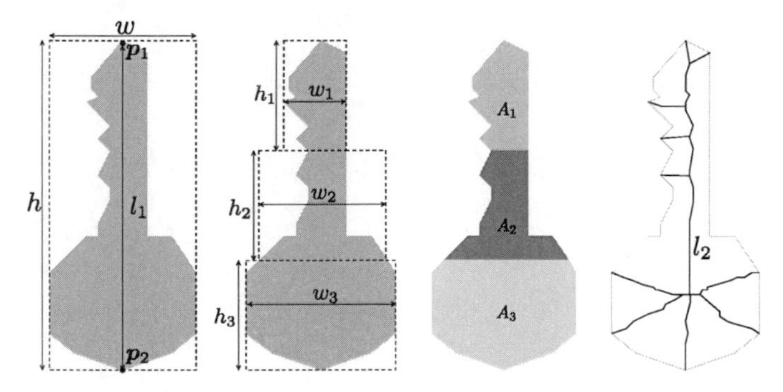

Figure 3.7: Shape bounding box and equally high sub-boxes ($h_1 = h_2 = h_3$) used for feature extraction; A_1, A_2 and A_3 are the areas of the top, middle, and bottom sub-objects, respectively.

Prior to feature extraction, the orientation of each shape is adjusted by rotating it to the point that the straight line connecting its two maximally distant contour points becomes vertical and the majority of contour points lie on the left side of this line (See Figure 3.7). If the number of contour points on both sides of the line $\mathbf{p}_1\mathbf{p}_2$ is the same, the orientation is adjusted so that the straight line connecting its two maximally distant contour points becomes vertical and the majority of contour points lie on the upper $h/2$ side. If the shape is a star-like or circle-like shape, one straight line connecting its two maximally distant contour points is selected and the shape rotated so that the straight line becomes vertical.

After rotation, an object shape is described by a 10-dimensional feature vector D'. For this, similar to the open curve descriptor in Section 3.2, the bounding box of the whole shape as well as the three equally high sub-boxes are used for feature generation (See Figure 3.7). For feature generation, the first feature f'_1 and the last feature f'_{10} of the feature vector express the length of a shape contour and the length of a shape skeleton l_2, respectively. Please be noticed that the shape skeleton here is generated by the fast thinning algorithm in [ZS84] without any pre- and post-processing steps. The detailed description and discussion regarding shape skeletons are addressed in Section 4.3. The remaining features are computed as follows:

$$\begin{array}{llll} f_2' = \frac{h}{w} & f_3' = \frac{h_1}{w_1} & f_4' = \frac{h_2}{w_2} & f_5' = \frac{h_3}{w_3} \\ f_6' = \frac{A_3}{A_1} & f_7' = \frac{A_2}{A_1} & f_8' = A_1 + A_2 + A_3 & f_9' = l_1 \end{array} \quad . \tag{3.5}$$

Subsequently, two feature normalisation steps are performed. First, in order to ensure scale invariance, the non-ratio elements (f_1', f_8', f_9' and f_{10}') of the feature vector are divided by a half of the bounding box perimeter. Second, all the feature values are linearly scaled to the range $(0, 1]$ with the following equation:

$$D = \frac{D' - \min\{f_1', \cdots, f_{10}'\} + 1}{\max\{f_1', \cdots, f_{10}'\} - \min\{f_1', \cdots, f_{10}'\} + 1} \quad . \tag{3.6}$$

In order to avoid the situation that $D = 0$ and zero denominator, a value 1 is added to both numerator and denominator. The scaling is needed for the Support Vector Machine (SVM) [SV99] applied in the classification step in Section 5.2.2. The main advantage of scaling is to avoid attributes in greater numeric ranges dominating those in smaller numeric ranges. Another advantage is to avoid numerical difficulties during the calculation. Because kernel values usually depend on the inner products of feature vectors (e.g., the linear kernel and the polynomial kernel), large attribute values might cause numerical problems. After the normalisations, a shape can be represented by a feature vector D with 10 normalised feature values:

$$D = (f_1, f_2, \cdots, f_{10})^{\mathsf{T}} \quad . \tag{3.7}$$

3.4 Summary

In this chapter, two simple and intuitive descriptors using coarse-grained features are proposed. There are several advantages to both descriptors. Specifically, both descriptors can be easily fused with other meaningful descriptors like SC [BMP02], etc. This leads to a significant increase in descriptive power of the original descriptors without adding to much computation complexity. Moreover, several normalisation processes are applied to ensure the rotation, scaling and transformation invariance of the proposed descriptors. In addition, the proposed two descriptors are designed to complement each other. In particular, compared to the descriptor in Section 3.3, the descriptor in Section 3.2 has finer-grained features. However, it cannot be applied to shapes with specific topologies such as circles. The main reason is that contour partition points cannot be properly generated. In such a case, the descriptor Section 3.3 can be applied to fix this problem.

In Chapter 5.2, the matching algorithms for both descriptors are proposed and analysed. The algorithm can easily adapt to a concrete application domain or a dataset by learning weights assigned to different dimensions of the feature space. Their superior performances have been proven by the experiments in Section 6.2.

Chapter 4

Shape Representation with Fine-grained Features

Fine-grained shape descriptors mainly capture the shape features from both global and local object deformations. The main motivation is that many shapes have a similar global topology, but they are mixed up with deformations in some local regions (Figure 4.1). Moreover, shapes with intra-class variations are normally non-linear deformations which is more challenging, especially when multiple deformations are combined simultaneously with intra-class variations and geometric transformations. Therefore, it is critical to extract shape descriptors which are representative, discriminative and robust for fine-grained shape matching and retrieval. The desired descriptors should not only tolerate geometric transformations, non-linear deformations and intra-class variations, but also be efficient to discriminate shapes from different classes [Yan+16e]. Therefore, two robust shape descriptors are proposed in

Figure 4.1: Sample shapes with deformations in some local regions.

this chapter based on fine-grained shape features. Similar to the coarse-grained shape descriptors in Chapter 3, the proposed descriptors in this chapter are also classified into contour- and region-based methods based on the way of feature generation. In Section 5.3, the correlated matching methods are introduced.

4.1 Techniques for Fine-grained Feature Generation

Fine-grained shape descriptors are normally classified into two types: contour- and region-based methods. A detailed discussion and review of those methods can be found in [YIJ08].

For contour-based methods, one of the most common methods used is SC [BMP02]. SC features contour points with histograms in which bins are uniformly divided log-polar space. Since this method does not know which contour point is useful for matching, it requires a large number of contour points to achieve accurate correspondences and alignments. Sharon *et al.* [SM06] proposed the figure print of a shape. It is generated by a series of conformal maps, starting from mapping the object to a unit circle in the complex plane, then from the boundary of the object to the exterior of the circle, so that the final boundary is a diffeomorphism from the unit circle to itself. However, this method requires high computational complexity for feature generation. Moreover, as the shape distance is computed by the shortest path between geodesic connections, it is sensitive to noise and small perturbations. In contrast to these approaches, Maney *et al.* [Man+06] used integral invariants to describe shapes with similar invariant properties as their differential counterparts. An advantage of such a structural invariant approach is the ability to handle occlusions and possibility of partial matching in shapes. With this motivation, several invariant-based shape representation methods have been proposed [HS15; Yan+16e; JB14]. There are four types of commonly used invariants: (1) Algebraic invariants [SC00], (2) Geometric invariants [SN96], (3) Differential invariants [Cao+12] and (4) Integral invariants [HS15; JB14]. A broad review of those types of invariants used for shape representation is given in [Cao+12].

Recently, in order to capture more fine-grained features from a shape contour, some methods make use of both local and global shape contour features to derive a hierarchical descriptor. McNeill *et al.* [MV06] introduced a hierarchical procrustes method for shape matching based on shape contour segmentation. The hierarchical representation avoids the problems associated with pure global or local descriptors. Xu *et al.* [XLT09] established the contour flexibility method to extract both global and local features that depicts the deformable potential at contour points. Felzenszwalb *et al.* [FS07] presented the shape tree method which segments a curve into two halves in the middle, and the two halves are further segmented into respective halves. Raftopoulos *et al.* [RK11] proposed a method based on global-local transformation to represent shape curvature with view area representation, which is robust to noise. Bai *et al.* [BRW14] introduced the shape vocabulary representation with the idea of a bag of words, where the shape contours are segmented into fragments and represented as words of shape contours in different scales. Although most of those descriptors are invariant to shape rotation, translation and scaling (see Figure 2.12), they need to sample many contour points to precisely represent its characteristics. The main reason is that since those methods do not know which contour point is useful for matching, they need to use a large number of contour points to achieve accurate correspondences and alignments. Thus, all of the above-mentioned descriptors incur high computational complexity. In order to solve this problem, a contour-based method is proposed in Section 4.2 using only a limited number of interesting contour points. The proposed shape representation method can efficiently reduce the matching complexity while keeping the crucial shape geometry and topology.

Region-based shape descriptors take advantage of the information from inside of a shape. One of the most commonly used method is skeleton. Skeleton is an important shape descriptor for deformable object matching since it integrates both geometrical and topological features of an object. Skeleton-based descriptors usually lead to a better performance than contour-based shape descriptors in the presence of partial occlusion and articulation of parts [SK05]. This is because skeletons have a notion of both the interior and exterior of the shape, and are useful for finding the intuitive correspondence of deformable shapes.

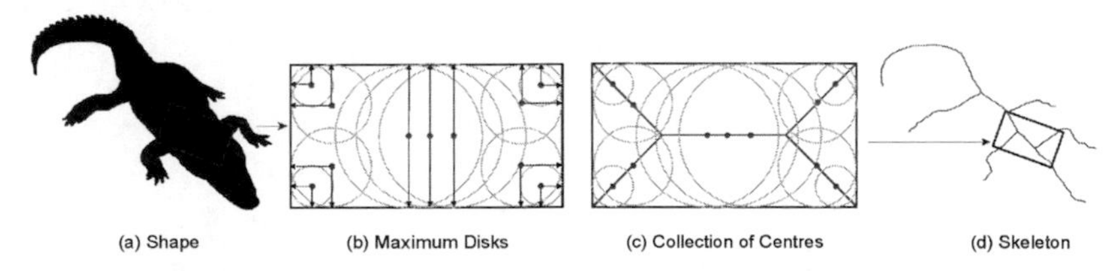

Figure 4.2: An overview of the skeletonisation process to convert a given shape (a) into a skeleton (d). (b) and (c) visually illustrate the skeleton extraction process, where the skeleton (red line) of a shape (rectangle) is generated by collecting the centres (red dots) of all discs (green dotted circles) that touch the boundary of the shape in two or more different locations (dotted arrows).

In order to generate proper skeletons for object matching, several skeletonisation methods have been developed [LLS92; MR94; Gor+06; DDS03]. One typical approach is the Max-Disk Model [CLS03] which continuously collects the centre points of maximal tangent disks that touch the object boundary in two or more locations. Figure 4.2 shows an overview of this process to convert a given shape (a) into a skeleton (d). Particularly, a skeleton is defined as a connected set of medial lines along the limbs of its shape [Dav04]. From a technical point of view, such a skeleton is extracted by continuously collecting centre points of maximal tangent disks touching the object boundary in two or more locations, as shown in Figure 4.2 (b) and (c). The centre point of a maximal tangent disk is referred to as a skeleton point. The sequence of connected skeleton points is called a skeleton branch. A skeleton point having only one adjacent point is an endpoint (the skeleton endpoint). A skeleton point having three or more adjacent points is a junction point.

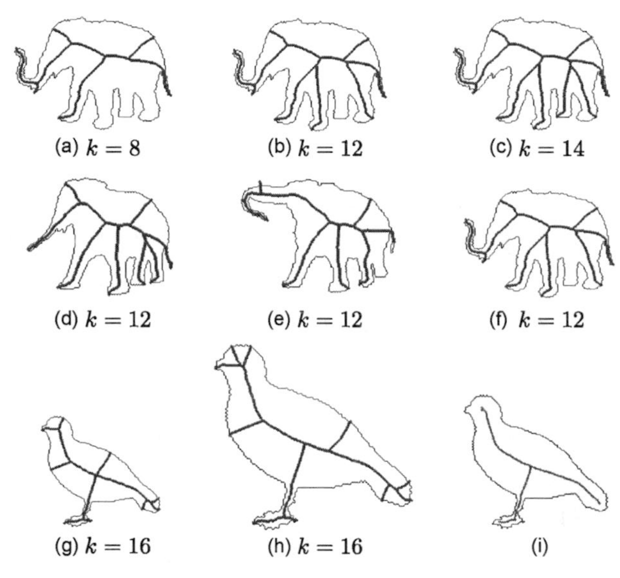

Figure 4.3: Skeletons on the same elephant shape with different DCE stop parameters k (first row), skeletons on different shapes with the same DCE stop parameter k (second row), and skeletons on the shape with different scales and the same stop parameter ((g) and (h) in the third row), and a pruned skeleton obtained by the method in [CD15] ((i) in the third row).

However, the skeleton obtained by this approach is sensitive to small changes and noises in the object boundary [OK95; SP14]. The intrinsic reason is that a small protrusion on the boundary may result in a large skeleton branch. To solve this problem, Choi *et al.* [CLS03; CD15; TW02] proposed algorithms to detect the skeleton in a distance map of the boundary points. Figure 4.3 (i) shows a skeleton obtained by the method in [CD15]. Although these methods can preserve some visual parts of a shape, some significant parts are missing. Therefore, they cannot guarantee the completeness of a skeleton. To overcome this, Bai and Latecki present significance measures for skeleton pruning associated with DCE [BLL07] or Bending Potential Ratio (BPR) [She+11]. Both methods decide whether or not a skeletal branch should be pruned by evaluating the contribution of its corresponding boundary segment to the overall shape. However, these methods require manual intervention to stop the evaluation and produce visually pleasing skeletons. For example, in Figure 4.3,

DCE requires a proper stop parameter k to calibrate the pruning power. However, different stop parameters for the same object (the first row in Figure 4.3) or the same parameter for different objects (the second row in Figure 4.3) lead to visually different skeletons in which some important parts are missing (legs in Figure 4.3 (a), (b), (e) and (f)). Furthermore, even if the best stop parameter is found, skeletons of the same object sometimes differ if the scale is changed (Figure 4.3 (g) and (h)). This is because the vanishing of shape parts is unavoidable when the resolution decreases [BRB01]. Therefore, fixing k for skeleton pruning is not a proper solution for all objects. In contrast, in Section 4.3, a shape is represented by a hierarchical skeleton which is a collection of skeletons obtained by all the stop parameters. This not only eliminates the necessity of manually tuning a stop parameter, but also preserves both the global and local properties of a shape.

Recently, researchers begin to combine both contour- and region-based shape features for robust shape matching and classification. Bai *et al.* [BLT09] proposed to integrate shape contour and skeleton for object classification. This method can also be extended to incorporate other shape features like SC [BMP02] and Inner-Distance (ID) [LJ07], etc. Shen *et al.* [She+16b] proposed to recognise shapes by a new shape descriptor, Skeleton-associated Shape Context (SSC), which captures the features of a contour fragment associated with skeletal information. Benefiting from the association, this descriptor provides the complementary geometric information from both contour and skeleton parts, including the spatial distribution and the thickness change along the shape part. Motivated by those methods, in Section 5.4, several shape matching methods based on the integrated coarse- and fine-grained shape features are introduced. Moreover, those methods are evaluated by the experiments in Section 6.4.

4.2 Contour-based Method

In shape-based object matching, it is important how to fuse similarities between points on a shape contour and the ones on another contour into the overall similarity. However, as introduced in Section 4.1, since most contour points are involved for possible matchings without taking into account the usefulness of each point, it causes high computational costs for point matching. In order to solve this problem, the proposed contour-based method represents a shape with only a small number of interesting points. Each interesting point is defined as a contour point that represents a rigid region of a shape. More specifically, shapes are normally composed of different regions (Figure 4.4 (a)) and some regions are likely to be deformed when changes occur in the shape. However, some regions are resilient against shape deformations like the bird's head, bone ends or the handle of a hammer in Figure 4.4. Such a region is regarded as rigid, and modelled as a region which is deviated from the overall shape trend. As interesting points are mainly detected in the rigid regions, they are robust to shape deformation (Figure 4.4 (b)).

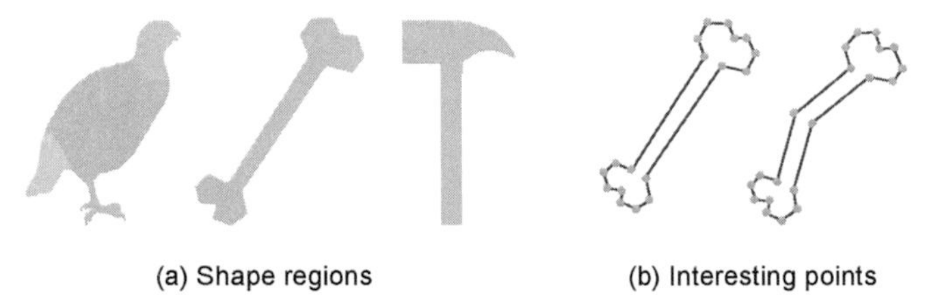

(a) Shape regions (b) Interesting points

Figure 4.4: Shape regions and shape interesting points.

With the detected interesting points, a simple and highly discriminative point descriptor, namely Point Context, is introduced to represent the geometrical and topological location of an interesting point. In this section, the method which generates robust interesting points along the shape boundary is introduced first. After that, a point context descriptor is introduced and analysed. Based on interesting points and point context descriptors, a shape is finally represented.

Interesting Points: In this part, the problem of selecting a set of interesting points $\{\mathbf{p}_i\}_{i=1}^N$

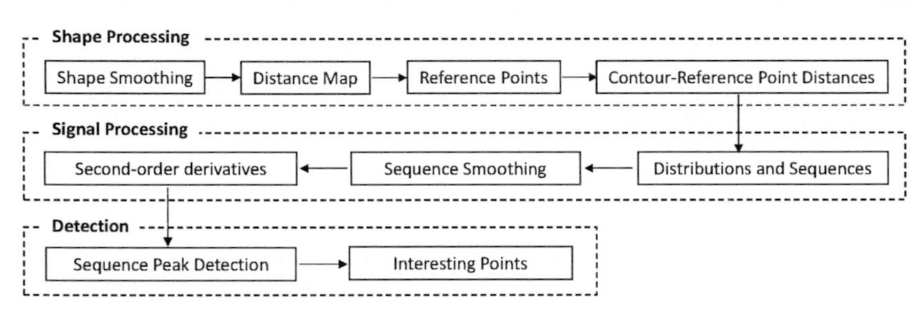

Figure 4.5: A pipeline of the proposed interesting point detection method.

from a given shape D is considered. It is assumed that distinctive contours like the legs or the tail of an elephant are characterised by a high curvature towards the overall shape trend. Based on this idea, as illustrated in Figure 4.5, the distance between each single contour point and its closest reference point is computed. Here, a reference point is a point that lies inside the shape and characterised by the highest distance to the contour. By arranging these values sequentially, a sequence $\mathbf{s}$ is generated where interest points characterised by high curvatures are detected as peaks. The detailed process of the proposed method (Figure 4.5) is introduced below.

Since noises on a contour could have adverse influence on interesting point detection, a polygonisation process is firstly performed to suppress noises without removing significant parts of the contour. For this purpose, the well-known *Douglas-Peucker* technique [Ebi02] is recursively applied to the object's contour. Then, the contour is converted into a polygon P.

Figure 4.6: Polygonisation process of a bird using Douglas-Peucker [Ebi02].

Having P, the reference points $\mathbf{x}_i$ (residing inside the shape) can be detected by utilising a Fast Marching Method (FMM) [HF07]. The FMM method is employed for reference point detection since it is convenient to use while delivering a high accuracy result. The approach is initialised with the whole polygon (interpreted as a set of points) leading to a distance map V. This map provides the distance from each pixel to the closest contour point on P, as shown in Figure 4.7.

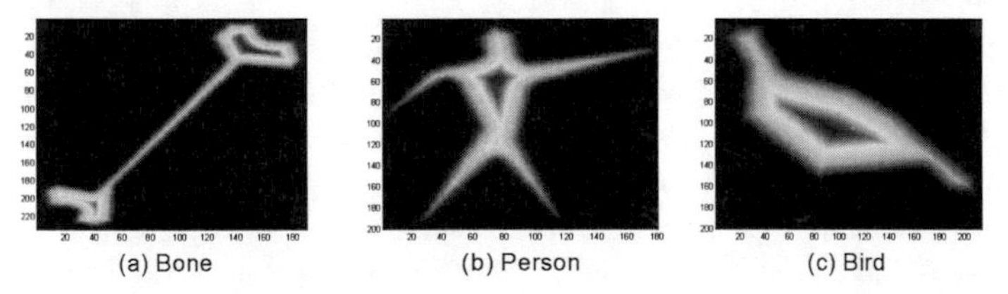

(a) Bone (b) Person (c) Bird

Figure 4.7: The distance map V of a bone, a person and a bird.

Reference points are then discovered at maximum value locations in V. To find reference points that appropriately cover the whole shape, V is iteratively masked and regions are detected from which reference points are extracted. This masking is carried out using the following dynamically adapted threshold $\mu^{(BG)}$:

$$\mu^{(BG)} = \phi(V(D)) - 2 \cdot \tau(V(D)) \quad . \tag{4.1}$$

where $\phi(\cdot)$ returns the maximum value inside V restricted to the area of D and $\tau(\cdot)$ indicates the standard derivation. Here, another option is to use $\phi(\cdot)$ mean instead of maximum as maximum is an unstable statistic compared to mean. However, it leads to less stability in interesting point detection and worse shape retrieval results in our preliminary experiments. The main reason is that reference points should be discovered at the maximum value in the most prominent regions in V. If $\mu^{(BG)}$ is assigned with the $\phi(\cdot)$ mean value, some irrelevant regions could be involved due to the low threshold $\mu^{(BG)}$. By masking V with $\mu^{(BG)}$ in Equation (4.1), the most prominent regions can be emphasised by removing irrelevant regions. In particular, the former regions are clustered into disjoint regions E_i. For each region, a reference point is determined as a weighted centroid where the weight of a contour point is its value in V. In the case where E_i may include multiple sub-regions, a set of reference points corresponding to these sub-regions are obtained by iteratively updating $\mu^{(BG)}$ and considering the detected reference points as seed point for further FMM. This iteration is necessary to find reference points for complex shapes which have multiple rigid regions. The algorithm terminates if the current iteration returns an empty set of new references. The detected reference point set is denoted as Ψ'.

Based on the collected reference points, the following two types of filtering are conducted to eliminate each meaningless reference point. First, none of the reference points are allowed to be too close to the shape's centroid, which is ensured by the threshold $\mu^{(cog)}$. If only one reference point $\mathbf{x}_i \in \Psi'$ is violating this constraint, all reference points are discarded except for the responsible one. Second, in presence of multiple reference points, the contour has to be split into sub-parts. During this separation, the algorithm monitors that each $\mathbf{x}_i$ is only assigned to one region. If one or two regions are assigned, the point will be removed. This strategy can ensure $\mathbf{x}_i$ only represents one region of the shape. Moreover, it is easier to sample the contour points sequentially in the next steps. The reference point set that passed the above-mentioned filtering is denoted as Ψ.

Using Ψ, the aim is to compute the distance between each contour point on the shape and its closest reference point. This is done by FMM which uses Ψ as seed points and outputs $V^{(final)}$. By sequentially aggregating values corresponding to contour points in $V^{(final)}$, a sequence $\mathbf{s}$ is obtained where each element represents the distance between a contour point and its nearest reference point. Based on $\mathbf{s}$, shape interesting points can be extracted which have a high influence on the perceptual appearance of the shape. Figure 4.8 shows both the $V^{(final)}$ values of the contour and their signal plots. Please notice that the bone structure yields two reference points so that the contour has been separated into two parts (first and second column).

It is obvious that the noise-to-signal ratio degrades the peak detection with imprecise interesting point detection. Thus, $\mathbf{s}$ is smoothed using a Low Pass Filter (LPF) based on a Fast Fourier Transform (FFT). Before LPF, the sequence is padded at the start and at the end to alleviate border artifacts. This padding is needed for shapes that are separated into several parts like the bone in Figure 4.8. For such a shape, without padding, contour points at boundaries of parts could be falsely detected as interesting. Figure 4.9 shows sequences

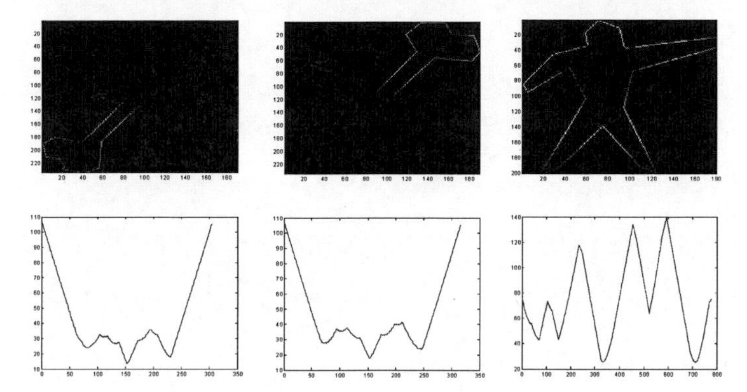

Figure 4.8: Example of sequences representing distances between contour points and their closest reference points. The colours on the contour line encode the distance from a pixel to its closest reference point.

smoothed by applying LPF to the ones in Figure 4.8. Here, the LPF is established with a Gaussian coefficient mask that is applied to the zero-shifted frequency domain.

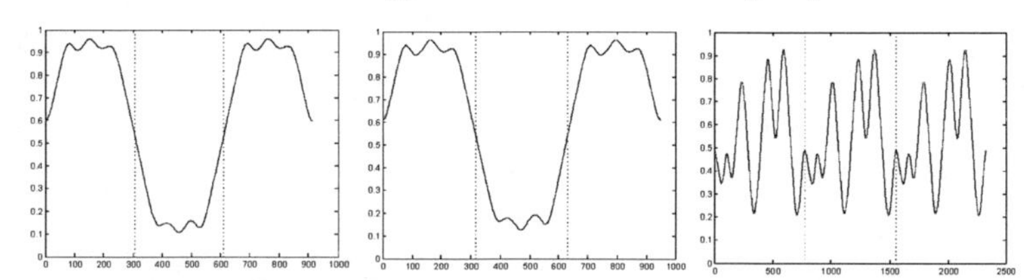

Figure 4.9: Sequences obtained by smoothing sequences in Figure 4.8 (each smoothed sequence corresponds to the middle part of the sequence).

Finally, interesting points are detected as peaks in the smoothed sequence $\hat{\mathbf{s}}$. Each peak is identified as a point where the first-order derivative ($\hat{\mathbf{s}}'$) of $\hat{\mathbf{s}}$ is zero, and the second-order ($\hat{\mathbf{s}}''$) is positive or negative. Figure 4.10 shows the second-order derivatives for the signal plots in Figure 4.9. Please note that the computation of the second-order derivative only considers a signed binary version of the first-order derivative (sign($\hat{\mathbf{s}}'$), where sign($\cdot$) returns 0 if $\hat{\mathbf{s}}'_i = 0$, 1 if $\hat{\mathbf{s}}'_i > 0$ and -1 otherwise).

As shown in Figure 4.11, the naive approach which locates interesting points based on $\hat{\mathbf{s}}'' = 0$ may cause several false-positives that multiple interesting points are located within a very short distance. To avoid such false-positives, a distance threshold $\mu^{(sdiff)}$ is applied to $\hat{\mathbf{s}}''$. If the distance between adjacent peaks is below $\mu^{(sdiff)}$ coupled with a low height

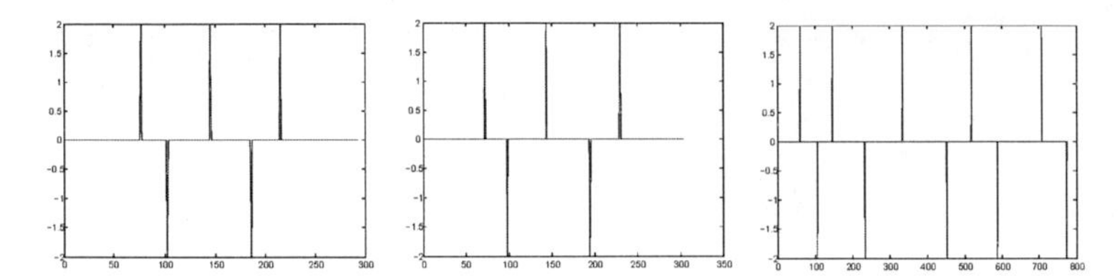

Figure 4.10: Second-order derivatives that correspond to the signals shown in Figure 4.9.

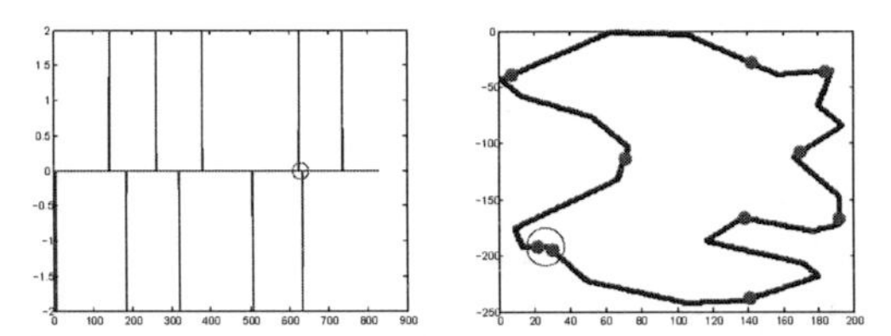

Figure 4.11: A situation where the sigma of the Gaussian filter has not been chosen appropriately. It is obvious that these artifacts can be easily determined by analysing the second-order derivative. The red circles indicate the problem.

difference (taken from $\hat{\mathbf{s}}$), the power of the LPF is dynamically decreased. With the lower smoothing power, the procedure is repeated with the original sequence $\mathbf{s}$ until the peak distance constraint is fulfilled. Based on these validated peaks, interesting points can be extracted which highly indicate shape characteristics. Figure 4.12 shows the final interesting points of a bone, a person and a bird.

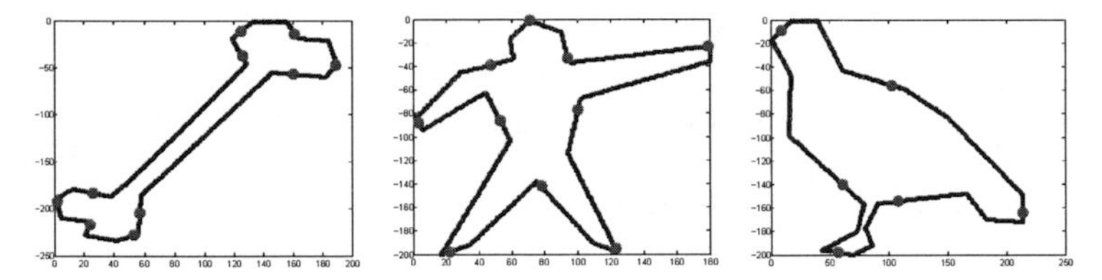

Figure 4.12: Interesting points of a bone, a bird and a person.

Point Context: Inspired by [ML11a; Lu+09b], the point context descriptor is proposed representing each interesting point $\mathbf{p}_i, (i = 1, 2, \cdots, N)$ based on its geometrical and topological location. It is calculated by a set of vectors originating from $\mathbf{p}_i$ to all other sample points on a shape contour. These vectors express the configuration of the entire shape relative to $\mathbf{p}_i$. Theoretically, instead of contour sample points, the feature vectors can also be formed by only using interesting points. However, this strategy is not employed for the following reasons: (1) It will lose some coarse- and fine-grained features since the number of interesting points is limited. (2) It will reduce the robustness of point context since the proposed method may fail to locate interesting points on some characteristic contours. This causes the dramatic change of the point context. (3) The major computation time derives from matching interesting points rather than the vector computation in point context extraction. Therefore, the difference of computation time between the two strategies can be ignored.

Let $\mathbf{P}$ denote a sequences of interesting points $\mathbf{P} = \{\mathbf{p}_1, \cdots \mathbf{p}_N\}$ and $\mathbf{Q}$ denotes a finite number of contour sample points $\mathbf{Q} = \{\mathbf{q}_1, \cdots, \mathbf{q}_M\}$, $\mathbf{P} \notin \mathbf{Q}$. All points in $\mathbf{P}$ and $\mathbf{Q}$ are represented by their coordinate locations. Points in $\mathbf{Q}$ are ordered clockwise along the shape contour. For $\mathbf{p}_i$, two vectors are computed, one presenting the distance of $\mathbf{p}_i$ to each $\mathbf{q}_j \in \mathbf{Q}(j = 1, \cdots, M)$, and the second representing the orientation of the vector from $\mathbf{p}_i$ to $\mathbf{q}_j$. A distance $\mathcal{D}_{\mathbf{p}_i}(j)$ from $\mathbf{p}_i$ to $\mathbf{q}_j$ is defined as Euclidean distance in the log space:

$$\mathcal{D}_{\mathbf{p}_i}(j) = \log(1 + \|\overrightarrow{\mathbf{p}_i} - \overrightarrow{\mathbf{q}_j}\|^2) \quad . \tag{4.2}$$

In order to avoid the divergence of log, one is added to the Euclidean distance. An orientation $\Theta_{\mathbf{p}_i}(j)$ from $\mathbf{p}_i$ to $\mathbf{q}_j$ is defined as the orientation of vector $\overrightarrow{\mathbf{p}_i} - \overrightarrow{\mathbf{q}_j}$:

$$\Theta_{\mathbf{p}_i}(j) = \mathrm{atan2}(\overrightarrow{\mathbf{p}_i} - \overrightarrow{\mathbf{q}_j}) \quad . \tag{4.3}$$

where atan2 stands for the four quadrant inverse tangent which can ensure $\Theta_{\mathbf{p}_i}(j) \in [-\pi, \pi]$. Together with the distances, a single interesting point $\mathbf{p}_i$ is encoded as two M-dimensional vectors $\mathcal{D}_{\mathbf{p}_i}$ and $\Theta_{\mathbf{p}_i}$.

The proposed point descriptor is different from the method in [BMP02]. Firstly, the proposed point descriptor only considers the feature vectors on the basis of interesting points instead of uniformly or randomly selecting sample points. This strategy can reduce the mismatches and computational complexity conspicuously. Secondly, the proposed point descriptor is naturally translation and scaling invariant since the distance between point contexts is computed by normalising $\mathcal{D}_{\mathbf{p}_i}(j)$ and $\Theta_{\mathbf{p}_i}(j)$ (see Equation (5.12) and Equation (5.13) in Section 5.3.1). In addition, the point context features are generated by the Euclidean distance and the four quadrant inverse tangent methods; their values remain the same even a shape is rotated. Thus, the proposed descriptor is also rotation invariant. On the contrary, the approach in [BMP02] is not intrinsically rotation invariant because each point is characterised by the tangent angle which is ineffective for those points for which no reliable tangent can be computed.

Shape Representation: Finally, given an arbitrary shape D, its contour ∂D can be represented with the locations as well as the distance and orientation vectors of all contour interesting points:

$$\partial D = \{\mathbf{p}_i, \mathcal{D}_{\mathbf{p}_i}, \Theta_{\mathbf{p}_i}\} \quad . \tag{4.4}$$

Here, the rotation property of Equation (4.4) is briefly discussed. Theoretically, for a single interesting point, the proposed point context descriptor is rotation invariant since it employs methods like Euclidean distance and four quadrant inverse tangent which can stabilise the point context features even if a shape is rotated. Considering a shape with multiple interesting points, the proposed global shape descriptor in Equation (4.4) is not completely invariant to rotation since the order of interesting points could be changed if a shape is rotated. This problem can be easily solved by some shape preprocessing methods [Yan+15b]. However, as stated in [BMP02], complete invariance sometimes impedes the recognition performance. Thus, the complete rotation invariance of Equation (4.4) is applied based on different applications.

4.3 Region-based Method

For region-based shape representation, skeleton is an important shape descriptor for deformable object matching since it integrates both geometrical and topological features of an object. However, as introduced in Section 4.1, a skeleton is sensitive to the deformation of an object's boundary because little variation or noise of the boundary often generates redundant skeleton branches that may seriously disturb the topology of the skeleton [She+11; HPF14; CKK15]. Furthermore, a large number of skeleton branches may cause the overfitting problem and high computation complexity. Although skeleton pruning [BLL07; She+11] approaches can remove the inaccurate or redundant branches while preserving the essential topology, they normally require manual intervention to produce visually pleasing skeletons. Moreover, the performance of skeleton-based matching highly depends on the quality and completeness of skeletons.

Figure 4.13: Examples of shapes that are perceptually similar to the original one, irrespective of fine-grained noises and deformations.

To overcome these problems, a hierarchical skeleton-based object representation method is proposed in this section. A hierarchical skeleton is a set of skeletons that represent an object at different levels. More specifically, during the skeleton pruning process, all the pruned branches are stored until the skeleton is pruned to the simplest form. These branches are reused to construct the hierarchical skeleton which is favourable for the following reasons: First, it does not need any manual intervention since it is generated by considering a set of skeletons rather than a single one. Second, a hierarchical skeleton captures geometric and topological features at different levels along with skeleton pruning. Fine levels feature the small object deformation while skeletons at coarse levels capture global shape deformations. This design is based on the fact that shapes (e.g. the four shapes on the right side of the arrow in Figure 4.13) reconstructed with the same skeleton topology are still perceptually similar to the original (the triangle on the left side of the arrow in Figure 4.13) even though there are some fine-grained noises and deformations.

The third advantage of a hierarchical skeleton is that it can also provide additional information for improving the object matching accuracy. In particular, by looking into the skeleton pruning process, transitions of pruned skeletons from the same category are more similar than those from different ones. This is because skeletons from the same category have more similar branches and these branches on each level have similar effects on the possible skeleton reconstruction. This phenomenon is called skeleton evolution (See Figure 4.15). In Section 6.3.2, experiments show that adopting skeleton evolutions improves the performance of object matching. As the fourth advantage, the hierarchical skeleton is obtained along with the skeleton pruning process, requiring no extra computational cost. Lastly, by limiting levels of hierarchical skeletons, it can filter out skeleton branches which represent shape properties irrelevant to matching. This alleviates the overfitting problem.

The extraction of a hierarchical skeleton takes advantage of skeleton pruning which iteratively removes skeleton branches of visually insignificant shape parts based on the boundary abstraction method DCE introduced in Section 3.2. Figure 4.14 illustrates an overview of the pruning process: (1) Given a planar shape D (Figure 4.14 (a)), the initial skeleton $S_n(D)$ (Figure 4.14 (b)) is generated using the Max-Disk Model [CLS03] introduced in Section 4.1. n indicates the first iteration index of DCE and will be iteratively decremented until 3. k denotes one of these steps. (2) The shape boundary ∂D is regarded as the initial polygon P_n and will be simplified into the polygon P_k (blue solid line in Figure 4.14 (c)) using the polygon simplification method described below. (3) With P_k, $S_n(D)$ is pruned by removing all skeleton points $\mathbf{p}_s \in S_n(D)$ so that the generating points (the points of tangency between the shape contour and the max discs) of $\mathbf{p}_s$ are contained in the same boundary segment. A boundary segment here is defined as a part of the shape boundary which is

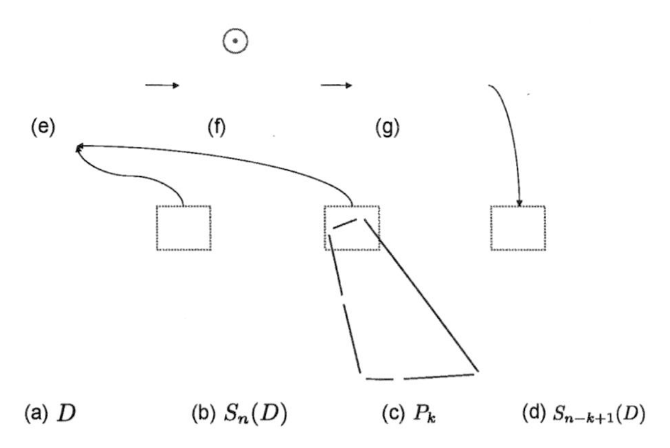

Figure 4.14: Illustration of the original shape D, initial skeleton $S_n(D)$, simplified polygon P_k and the pruned skeleton $S_{n-k+1}(D)$ generated by [BLL07] with $k = 10$. An intuitive description of the skeleton pruning process is illustrated in (e) (g) with the enlarged bird head.

approximated by the straight line (polygon partition) between two neighbouring vertices of P_k (red stars in Figure 4.14 (c)). Each pruned point $\mathbf{p}_s$ results from a boundary segment with respect to the polygon partition and therefore, $\mathbf{p}_s$ can be considered as an unimportant skeleton point and can be removed.

In order to describe the aforementioned processes more intuitively, the bird head (the part in the red-dotted rectangles) is enlarged in Figure 4.14 as an example. Specifically, the initial head skeleton and the head contour are illustrated together in (e). With vertices of P_k (the red points), the head contour is divided into three boundary segments (marked with three different colours in (f)). Here, a skeleton point $\mathbf{p}_s$ (the green point in (f)) is generated by a max-disc (the green circle) which has at least two touching points (generating points) to the shape contour. Since all the touching points of $\mathbf{p}_s$ are located in the same contour segment (with the pink colour), $\mathbf{p}_s$ is an unimportant skeleton point and can be removed. This removal process is iteratively applied until all unimportant skeleton points are erased. As a result, the skeleton in (f) is pruned and illustrated in (g).

The skeleton pruning described above is based on a simplified polygon P_k. The detailed DCE-based polygon simplification for generating P_k is introduced in Section 3.2. Essentially, DCE produces a sequence of simpler polygons $P = \{P_n, P_{n-1}, \cdots, P_3\}$ by iteratively removing polygon vertexes which have a low contribution calculated by Equation (3.1). By reordering polygons from simple to complex in the polygon simplification steps (i.e., $P = \{P_3, P_4, \cdots, P_n\}$), a hierarchical skeleton $S(D) = \{S_3, S_4, \cdots, S_n\}$ of a shape D can be constructed. As pruned skeletons are a by-product of the polygon simplification process, the total calculation cost stays the same to the skeleton pruning process. In order to reduce the computational cost of skeleton matching, $S(D)$ is built only using skeletons collected from the T_{min}th to T_{max}th iterations. Particularly, T_{max} is selected as the start step number based on the trade-off between computation time and matching results, $T_{max} > T_{min}$. With this, the DCE iteration can directly start from T_{max} and stops at T_{min}. An example of a hierarchical skeleton is shown in Figure 4.15 and the skeletons with $t \in [3, 15]$ are selected for matching. Setting $T_{min} = 3$ is based on the constraint that skeletons should have at least three endpoints and $T_{max} = 15$ is selected based on the preliminary experiments in [Yan+16b].

Figure 4.15: A hierarchical skeleton with DCE steps from $t = 3$ to 17 (top left to right down).

4.4 Summary

In this section, two fine-grained shape descriptors are proposed using contour- and region-based shape features. For the contour-based method, the main idea is to generate some interesting points on shape contours for correspondence matching. In order to fully capture the geometrical and topological locations of each interesting point, a simple and intuitive point context descriptor is proposed. This descriptor expresses the configuration of the entire shape relative to the interesting point. In addition, by normalising point contexts in terms of lengths and orientations, they become invariant to shape rotation and translation. Thanks to such point contexts integrated with interesting points, the proposed method features a high discriminative power while keeping low computational complexity. In order to fully utilise the properties of this descriptor, a high-order graph matching method is proposed in Section 5.3.1 which considers geometrical relations characterised by multiple interesting points. The robustness and discrimination power of this descriptor and its matching algorithm are validated by the experiments in Section 6.3.1. Moreover, the complexity analysis and runtime comparisons are also addressed in Section 6.3.1 to illustrate the efficiency of the proposed method for feature generation and shape matching.

For the region-based method, a hierarchical skeleton-based shape descriptor is introduced. The main idea is to reuse skeleton branches pruned in the skeleton pruning process since they contain some fine-grained geometric and topological information of the original shapes. Based on this, a hierarchical skeleton is generated by organising multiple skeletons obtained by the skeleton pruning process. Thus, a hierarchical skeleton captures geometric and topological features at different levels along with skeleton pruning. Moreover, it improves the stability of skeleton pruning without human interaction. In addition to the invariance properties on shape scaling, rotation and transformation, a hierarchical skeleton also provides additional information for accurate shape matching. In order to fully use these properties, a hierarchical skeleton matching algorithm is proposed in Section 5.3.2. The experiments in Section 6.3.2 demonstrate that the proposed method is significantly superior to most conventional shape descriptors and the single skeleton-based methods.

Chapter 5

Shape Matching using Coarse- and Fine-grained Features

In this chapter, several shape matching methods based on the proposed and fused shape descriptors are introduced. Before going into further discussion, it is important to clarify the meaning and the aim of shape matching. In some literary works [Man+06; BL08; BMP02], shape matching indicates a process of putting into correspondence different parts of two given shapes [Man+06]. For instance, establishing point correspondence among contours or skeletons, searching corresponding sub-regions between shapes, etc. Based on the correspondences between points or regions, a similarity (or dissimilarity) value can be calculated to distinguish shapes. However, for some shape descriptors, the corresponding-based shape matching cannot be properly applied due to the feature structure [MSJB15; Yan+14b] and time complexity constraints [ML11a; Lu+09b]. In such cases, a shape similarity (or dissimilarity) is calculated by vector distance methods [YK8; RTG00; RTG00; Bha46; Dan80] or feature statistics [ML11a; MSJB15]. No matter by which way the shape similarity (or dissimilarity) is calculated, the ultimate goal is to compare objects based on their similarity (or dissimilarity values). With these observations, shape matching in this thesis means a process of calculating similarity (or dissimilarity) between two given shapes based on their descriptors. This has obvious implications in shape classification for object recognition, content-based image retrieval, medical diagnosis, etc.

In the first section of this chapter, some existing shape matching methods are reviewed and discussed. After that, several shape matching algorithms are introduced in Section 5.2 and 5.3 based on the proposed shape descriptors in Chapter 3 and 4, respectively. Lastly, built on some existing and the proposed shape features, three fused shape descriptors are introduced in Section 5.4 for shape matching.

5.1 Techniques for Shape Matching

For coarse-grained features, shape matching is usually addressed by calculating similarities (or dissimilarities) between shapes without any corresponding process in which the correspondences between points or regions are searched. The most commonly used methods include Euclidean distance [Dan80], correlation [YK8], HI [RTG00], χ^2-statistics [RTG00] and Hellinger [Bha46], etc. Those methods normally treat a shape descriptor as a feature vector and then calculate the vector distances. The vector distances can be used as shape dissimilarity values. One drawback of those methods is that they consider each feature in a feature vector to have the same contribution to distinguish two given shapes. However, this is highly dependent on datasets. In practice, each feature in a feature vector may have a different power for discriminating given shapes in a dataset. Especially, discriminating properties of a feature space can be evaluated by some Discriminate Analysis (DA) methods like the Fisher Linear Discriminant Analysis [SM99]. If the discriminating power of each feature can be enhanced or reduced based on different shape datasets, the feature vector is more flexible and the discriminating power of a shape descriptor can be improved. With this motivation, in Section 5.2, two coarse-grained shape matching methods are proposed based on the correlated shape descriptors in Section 4.2. Moreover, the discriminating power of each feature in a feature space is controlled by certain parameters which can be assigned with a supervised optimisation strategy proposed in Section 5.2.2.3.

For fine-grained features, shapes are usually matched by searching correspondences on points or other elements of two given shapes. Comprehensive surveys of shape matching techniques with respect to correspondence can be found in [MS05; VK+11]. In contrast to coarse-grained shape matching methods, correspondence-based shape matching measures the similarity between shapes using element-to-element matching. Specifically, shape matching either uses the intrinsic statistical properties of the shapes or anatomical modelling and then corresponds the shape elements to compute a matching cost (or dissimilarity) [ZL04; JB14]. Shape matching for fine-grained shape features depends on the type of descriptor used [MD89]. For contour-based descriptors, a matching cost is calculated by searching correspondences between shape contour points [Bro+08]. Hausdorff distance [HKR93] is a classical matching method where the distance of two point sets is normally calculated by both the maximal and minimal distance between point pairs. Hence, this method is sensitive to noise and slight variations. Belongie *et al.* proposed a correspondence-based shape matching method using shape contexts [BMP02] wherein the matching of two shapes is done by matching their point histograms. For region-based features, shape matching techniques involve feature analysis based on graph matching [RWP05]. Bai *et al.* proposed a skeleton-based shape matching method which uses the Hungarian algorithm to find the best match of skeleton endpoints in terms of their geodesic paths [BL08].

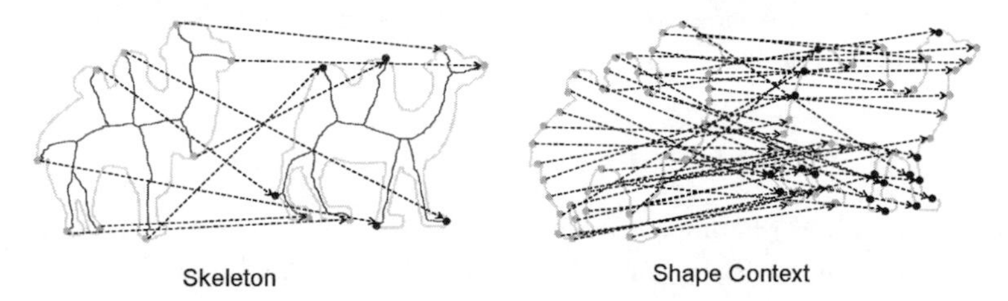

Figure 5.1: Point (green dots) correspondences between two camels with the skeleton-based matching method in [BL08] and the shape context-based matching method in [BMP02].

However, all these matching methods may suffer from the initial alignment problem due to boundary noise or shape symmetry. As shown in Figure 5.1, there are many mismatched points (marked in blue) in both the skeleton-based method [BL08] and the shape context-based method [BMP02]. To overcome this, one possible way is to avoid those points which are below a threshold of similarity value during the correspondence-based matching [AMO93; Kuh55; RTG00]. Another way is to jump over those similar points by adding dummy points in another object [BL08; Lat+07]. However, even these methods only consider the relationship between single points. This could lead to ambiguous matching because many different points may have similar descriptors [Fei+14; VK+11].

Compared to the above-mentioned methods, several matching approaches [ZS08; LH05] have been proposed by considering the geometric relations among multiple points. Leordeanu *et al.* proposed a spectral technique for matching problems using pairwise constraints [LH05] where the correspondence is established through preserving the structure similarity across two point sets. Nevertheless, this strategy could lead to a substantial loss of performance since every pair of points trivially defines a line which is repetitive and similar to each other. Therefore, instead of the singleton or pairwise, Zass *et al.* proposed to match points using hyper-graphs [ZS08] which are going beyond the pairwise. Specifically, each point set is modelled by a hyper-graph where the relations between points are represented by hyper-edges. A match between the point sets is then modelled as a hyper-graph matching problem. Due to the theoretical advance and empirical success, hyper-graph matching has attracted increasing attention and many methods have been proposed [LCL11; CK11; Duc+11; LLY10] and the references therein. However, it is unclear which potential functions are suitable for the matching tasks. Moreover, it is also interesting to explore the performances of different hyper-graph matching algorithms. Thus, with the proposed shape descriptors in Section 4.2 and 4.3, the singleton, pairwise and third-order potential functions are proposed for the interesting point and hierarchical skeleton matching problems. In addition, the matching performance between the employed and other hyper-graph matching algorithms are compared in Section section 6.3.1.3.

5.2 Shape Matching using Coarse-grained Features

In this section, based on the proposed shape descriptors in Section 3.2 and 3.3, the correlated shape matching methods are proposed. Both methods are proposed built on the feature vectors generated by coarse-grained shape features.

5.2.1 Shape Matching with Open Curve Similarities

This part indicates a shape matching algorithm based on the shape representation method in Section 3.2. Given two shapes $D^\star$ and $D^\diamond$, open curves of both shapes are firstly arranged in a clockwise way so that the shapes can be represented by ordered lists of feature vectors:

$$\begin{aligned} D^\star &= \{\mathbf{c}_1^\star, \mathbf{c}_2^\star, \cdots, \mathbf{c}_n^\star, \cdots, \mathbf{c}_{N_1}^\star\} \\ D^\diamond &= \{\mathbf{c}_1^\diamond, \mathbf{c}_2^\diamond, \cdots, \mathbf{c}_k^\diamond, \cdots, \mathbf{c}_{N_2}^\diamond\} \end{aligned} \quad . \tag{5.1}$$

To simplify further explanations, it is assumed that $N_1 \leq N_2$. Now, a dissimilarity measure is introduced for open curves belonging to different shapes $D^\star$ and $D^\diamond$:

$$d(\mathbf{c}_n^\star, \mathbf{c}_k^\diamond) = \frac{1}{M} \sum_{m=1}^{M} \frac{\sigma_m |f_{n,m}^\star - f_{k,m}^\diamond|}{\sum_{j=1}^{N_2} |f_{n,m}^\star - f_{j,m}^\diamond|} \quad . \tag{5.2}$$

where $M = 12$ is the dimensionality of the feature space (Equation (3.3) in Section 3.2) and σ_m is the weight for each feature achieved in an optimisation process explained at the end of this section. As one can see, the dissimilarity value between $\mathbf{c}_n^\star$ and $\mathbf{c}_k^\diamond$ does not only depend on these two open curves. All open curves of $D^\diamond$ are taken into consideration, whereby (Equation (5.2)) does not fulfill the symmetry property $d(\mathbf{c}_n^\star, \mathbf{c}_k^\diamond) \neq d(\mathbf{c}_k^\diamond, \mathbf{c}_n^\star)$. However, if the dissimilarity of $\mathbf{c}_n^\star$ to other open curves in $D^\diamond$ decreases, $d(\mathbf{c}_n^\star, \mathbf{c}_k^\diamond)$ increases. This behaviour is similar to human perception. For instance, a horse and a cat become, for humans, less similar to each other if a dog suddenly appears in the scene and chases away a stork [Tve77].

The values of the dissimilarity function (Equation (5.2)) belong to the range $d(\mathbf{c}_n^\star, \mathbf{c}_k^\diamond) \in [0, 1]$ which enables their easy conversion to similarity values:

$$s(\mathbf{c}_n^\star, \mathbf{c}_k^\diamond) = 1 - d(\mathbf{c}_n^\star, \mathbf{c}_k^\diamond) \quad . \tag{5.3}$$

Using Equation (5.3) a matrix of similarities between all open curves in $D^\star$ and $D^\diamond$ can be generated:

$$\mathbf{S}(D^\star, D^\diamond) = \begin{pmatrix} s(\mathbf{c}_1^\star, \mathbf{c}_1^\diamond) & \cdots & s(\mathbf{c}_1^\star, \mathbf{c}_{N_2}^\diamond) \\ \vdots & \vdots & \vdots \\ s(\mathbf{c}_{N_1}^\star, \mathbf{c}_1^\diamond) & \cdots & s(\mathbf{c}_{N_1}^\star, \mathbf{c}_{N_2}^\diamond) \end{pmatrix} \quad . \tag{5.4}$$

In order to find an optimum match of open curves from $D^\star$ to $D^\diamond$, the Hungarian algorithm [Kuh55] is finally applied on the matrix expressed in Equation (5.4). It should be noted that the Hungarian algorithm does not preserve the order of matched open curves. However, it does not influence the final score since the similarity of open curves is computed in the context of all other open curves. With this, even the order of open curves is changed, the correspondences remain the same. In addition, there is also a probability that the open curve numbers between two shapes are different and the generated matrix Equation (5.4) is not symmetric. In such a case, some additional rows with a constant value const can be added to Equation (5.4) so that the matrix becomes square. The constant value const is the average of all the other values in $\mathbf{S}(D^\star, D^\diamond)$. An intuition understanding for using the Hungarian algorithm is that a globally consistent one-to-one assignment of all open curves can be achieved, with possibly assigning some open curves to const (dummy curves). This means that each open curve in a shape can find one-to-one correspondence of a open curve in another shape, though some open curves could be skipped by assigning them to dummy curves. The resulting similarity values of the matched open curves can be denoted as $s_1, s_2, \ldots, s_{N_1}$ and the global similarity between the object contours $D^\star$ and $D^\diamond$ is calculated as follows:

$$s_{contour}(D^\star, D^\diamond) = \frac{1}{N_1} \sum_{n=1}^{N_1} s_n \quad . \tag{5.5}$$

It can be seen that in Equation (5.2) the dissimilarity value for two open curves depends on the weights σ_m, $m = 1, 2, \cdots, 12$. The weight of each feature expresses its importance for the overall similarity of two open curves. Setting the weights for a particular dataset gives the opportunity to adapt the matching algorithm to the application domain (context). In order to automatically estimate these weights, a Covariance Matrix Adaptation Evolution Strategy (CMA-ES) [HMK03] is applied. The algorithm is started with a configuration of equally distributed weights for all features and finds the optimum values for a certain dataset in an iterative process. In the practical realisation only a subset of each dataset is used for this optimisation.

5.2.2 Shape Matching and Classification with Supervised Optimisation

This part introduces the shape matching and classification methods based on the shape feature vector D in Section 3.3. In the first part, a shape matching method is introduced for object retrieval. After that, shape classification based on SVM is illustrated. In order to find proper parameters for classification, a supervised strategy is introduced. This strategy will also be used in the Section 5.3 and 5.4.

5.2.2.1 Shape Matching

For shape matching, a similarity measure between shapes is proposed. Assume $D^\star$ and D° are two feature vectors representing two object shapes. Based on Equation (3.7), $D^\star$ and D° are 10-dimensional feature vectors:

$$\begin{aligned} D^\star &= (f_1^\star, f_2^\star, \cdots, f_n^\star, \cdots, f_{10}^\star)^\mathsf{T} \\ D^\circ &= (f_1^\circ, f_2^\circ, \cdots, f_k^\circ, \cdots, f_{10}^\circ)^\mathsf{T} \end{aligned} \quad . \tag{5.6}$$

where $n,k \in [1,10]$. Now, a method for calculating dissimilarity between $D^\star$ and D° is introduced:

$$d'(D^\star, D^\circ) = \frac{1}{10}\sum_{m=1}^{10} \frac{\sigma_m |f_m^\star - f_m^\circ|}{|f_m^\star + f_m^\circ|} \quad . \tag{5.7}$$

where σ_m is the weight for each feature achieved in an optimisation process explained in Section 5.2.2.3. σ_m can be optimised to adapt the proposed feature vector to different datasets. Moreover, it helps the proposed feature to avoid the overfitting problem by applying a proper σ_m to different features. The proposed dissimilarity measure has been inspired by the Chi-Square kernel [Huz11], which comes from the Chi-Square distribution. Since the proposed shape descriptor contains a bag of features that are discretely distributed and Chi-Square kernel can effectively model the overlap among them. The values of the dissimilarity function (Equation (5.7)) belong to the range $d(D^\star, D^\circ) \in [0,1]$ which enables their easy conversion to similarity values:

$$s(D^\star, D^\circ) = 1 - d'(D^\star, D^\circ) \quad . \tag{5.8}$$

5.2.2.2 Shape Classification

In this part, shape classification using SVM [SV99] is introduced. SVM extracts a decision boundary between shapes of different classes based on the margin maximisation principle. Due to this principle, the generalisation error of the SVM is independent of the number of feature dimensions. Furthermore, a complex (non-linear) decision boundary can be extracted using a non-linear SVM. In this process, images in a high-dimensional feature space are mapped into a higher-dimensional feature space using a kernel trick. In order to properly classify shapes from different classes, the Radial Basis Function (RBF) is selected as kernel function for three reasons. Firstly, RBF kernel non-linearly maps samples into a higher dimensional space so that, unlike the linear kernel, it can handle the case in which the relation between class labels and attributes is non-linear. Secondly, The RBF kernel has less hyper parameters than the polynomial kernel which reduces the complexity of model selection. With RBF, there are only two parameters that need to be determined and they can be optimised by the proposed optimisation method in Section 5.2.2.3. Thirdly, as the number of instances is much larger than the number of features, the RBF kernel has fewer numerical difficulties and leads to a shorter training time.

In this work, a Multi-class Support Vector Machine (mSVM) is applied using its one-against-one (1vs1) version which works with a voting strategy. It uses a two-class SVM for each pair from a set of all considered classes $\{\omega_1, \omega_2, \cdots, \omega_N\}$. Thus, if there are N classes in total, $N(N-1)/2$ two-class classifiers have to be used. A sample pattern (query pattern) is classified using all these two-class SVMs. The final classification result is determined by counting to which class the sample pattern has been assigned most frequently.

5.2.2.3 Supervised Optimisation

The performance of matching and classification methods is heavily dependent on the choice of appropriate parameters. These parameters are mutually dependent and therefore need to be optimised simultaneously. In practice, parameters are normally selected and optimised manually based on the knowledge of experts. Obviously, this is an exhaustive and tedious process. In this section, an effective supervised optimisation strategy is proposed to automatically improve the accuracy of retrieval and classification methods (Figure 5.2).

Traditional optimisation methods use iterative strategies, which do not produce satisfactory results when applied to high dimensional problems. In contrast, heuristic methods are well suited for such optimisation problems where multiple parameters have to be optimised simultaneously. With this idea, a combination of two heuristic optimisation methods is proposed: Gradient Hill Climbing [RN09] integrated with Simulated Annealing [KGV83]. The

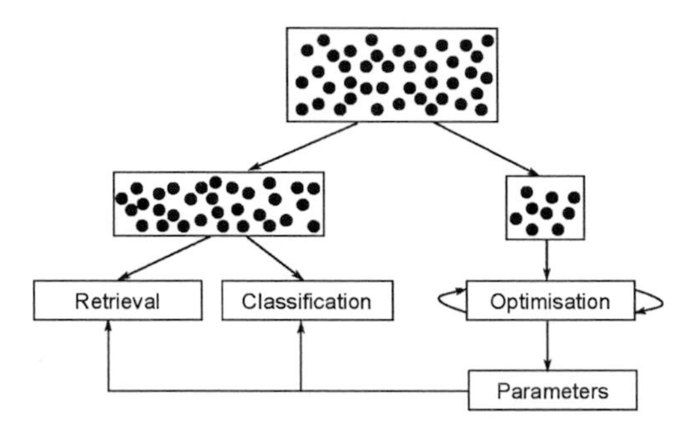

Figure 5.2: Pipeline of the proposed supervised optimisation method. Black points indicate shapes in a given dataset.

Gradient Hill Climbing method starts with randomly selected parameters. Then it changes single parameters iteratively to find a better set of parameters. A fitness function then evaluates whether the new set of parameters performs better or worse. The Simulated Annealing strategy impacts the degree of the changes. In later iterations, the changes to the parameters are becoming smaller. This strategy can efficiently reduce the computational complexity of the proposed optimisation method. By in- and decreasing all parameters separately with a specified magnitude that describes a convergent zero series, the gradient for maximum enhancement is computed. Adding this gradient to the previous parameters results in the parameters for the next iteration.

In order to use this heuristic strategy, a fitness function should be defined to evaluate results under a set of testing parameters. Specifically, for the task of shape classification in Section 5.2.2.2, the F-measure based on precision and recall values can be employed as the fitness function. For the task of shape retrieval in Section 5.2.2.1, the shape retrieval rate like the bull's-eye score [LLE00] can be used as a fitness function for evaluating the retrieval accuracy.

5.3 Shape Matching using Fine-grained Features

In this section, two matching algorithms are introduced for the proposed shape descriptors in Section 4.2 and 4.3, respectively. Instead of traditional singleton-based shape matching methods in Section 5.1, the matching methods in this section are applied by using higher-order matching strategies. Specifically, for the interesting point-based shape descriptor in Section 4.2, the matching method is applied using singleton and third-order potentials. For the hierarchical skeleton-based shape descriptor in Section 4.3, skeletons are matched using singleton and second-order potentials. The motivation for selecting and designing different orders and their correlated potential functions are also introduced in each subsection.

5.3.1 Contour Interesting Point Matching with High-order Graphs

Based on the properties of the shape descriptor in Section 4.2, the aim is to consider the geometric relations among multiple interesting points using high-order graph matching, which is an approach to match two graphs by extracting the correspondences of multiple nodes [TKR08]. This approach is adopted by considering nodes as interesting points described by point contexts. As shown in Figure 5.3(a), singleton point matching is a well-known assignment problem where the interesting point is matched with one point in another shape. For the pairwise matching (Figure 5.3(b)), it finds consistent correspondences between two pairs of interesting points by taking into consideration both how well their descriptors match and how similar their pairwise geometric relations are. For the high-order matching (mostly third-order, see Figure 5.3(c)), it considers the cost of matching three correspondences. More specifically, a triple of interesting points in a shape are matched with the one in another shape. With this observation, a high-order graph matching strategy is performed to improve the extraction of correspondences between interesting points.

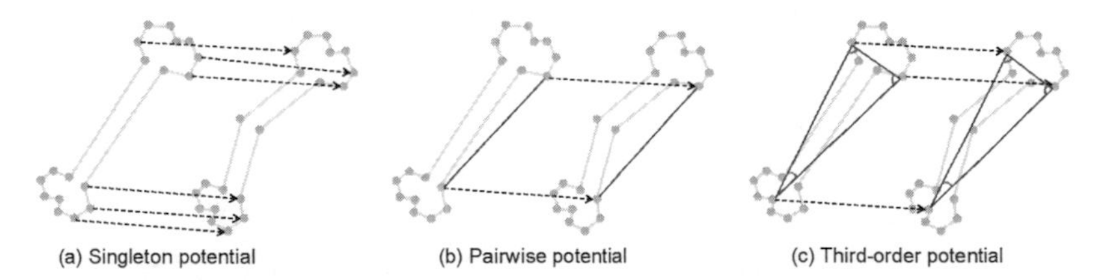

Figure 5.3: Different potentials for object matching.

In this section, shape matching is firstly formulated as high-order graph matching consisting of potential functions with different orders. Then, the definition of each potential function is introduced. Finally, a method which can efficiently find the optimal matching on high-order graphs is explained.

5.3.1.1 Formulation

Let $\mathbf{P}_1$ and $\mathbf{P}_2$ denote sets of interesting points from two shapes D_1 and D_2 respectively. $\mathbf{p}_i$ and $\mathbf{p}'_j$ denote a single interesting point in $\mathbf{P}_1$ and $\mathbf{P}_2$ respectively. $\mathbf{P} \triangleq \mathbf{P}_1 \times \mathbf{P}_2$ denotes the set of possible correspondences. The following boolean indicator is defined:

$$x_a = \begin{cases} 1 & \text{if } a = (\mathbf{P}_i, \mathbf{P}'_j) \in \mathbf{P} \text{ is a correspondence} \\ 0 & \text{otherwise} \end{cases} \quad . \tag{5.9}$$

In this definition, a basic constraint is that each point $\mathbf{p}_i$ in $\mathbf{P}_1$ is mapped to at most one point $\mathbf{p}'_j$ in $\mathbf{P}_2$, while for each point $\mathbf{p}'_j$ in $\mathbf{P}_2$ there is at most one point $\mathbf{p}_i$ in $\mathbf{P}_1$ mapped to it. Therefore, the following constraints are addressed:

$$\zeta = \{\mathbf{u} \in \{0,1\}^{\mathbf{P}_1 \times \mathbf{P}_2} \mid \sum_{\mathbf{p}_i \in \mathbf{P}_1} x_{\mathbf{p}_i, \mathbf{p}'_j} \leqslant 1, \sum_{\mathbf{p}'_j \in \mathbf{P}_2} x_{\mathbf{p}_i, \mathbf{p}'_j} \leqslant 1, \forall \mathbf{p}_i \in \mathbf{P}_1 \text{ and } \forall \mathbf{p}'_j \in \mathbf{P}_2\} \quad . \tag{5.10}$$

Inspired by [Duc+11; Zen+10], the proposed high-order (degree 3) matching formulation is formulated as the following optimisation problem:

$$\min_{\mathbf{u} \in \zeta} \{\mathcal{E}(\mathbf{u}|\theta) = \sum_{a \in \mathbf{P}} \theta_a x_a + \sum_{(a,b) \in \mathbf{P} \times \mathbf{P}} \theta_{ab} x_a x_b + \sum_{(a,b,c) \in \mathbf{P} \times \mathbf{P} \times \mathbf{P}} \theta_{abc} x_a x_b x_c\} \quad . \tag{5.11}$$

where θ is the whole set of matching costs that we consider and consists of the following three components: θ_a is the matching cost for each correspondence $a \in \mathbf{P}$ (Figure 5.3 (a)), θ_{ab} for a pair of correspondences $(a,b) \in \mathbf{P} \times \mathbf{P}$ (Figure 5.3 (b)), and θ_{abc} for a triplet of correspondences $(a,b,c) \in \mathbf{P} \times \mathbf{P} \times \mathbf{P}$ (Figure 5.3 (c)). Since the matching constraint in Equation (5.10) makes the optimisation problem in Equation (5.11) difficult to solve, a method that decomposes the problem in Equation (5.11) into several sub-problems is introduced in Section 5.3.1.3.

5.3.1.2 Potential Functions

In this section, only the first and the third order terms are considered for shape matching for the following reasons. Firstly, although singleton potential causes mis-matching of interesting points due to the lack of their topological relations, they still offer a major contribution to examine overall shape characteristics. Secondly, since the point contexts of interesting points are already considered in singleton potential, it is redundant to consider them for pairwise potentials. Thus, pairwise potentials can be defined based only on relative location relations of interesting points. However, the preliminary experiment showed that such pairwise potentials have low discriminative power, and many different pairs of points have similar descriptors. Hence, pairwise potentials will not be used in this part. Lastly, as discussed in [Duc+11], higher-order potentials make it possible to build more expressive features. This was also confirmed in the preliminary experiment, where triplets representing relative locations of three interesting points have a high discrimination power, even without considering their point contexts. This way, by assigning the similarity computation of point contexts only to singleton potential, the computational cost of high-order graph matching should be kept as low as possible.

The Singleton Potential: Next, the singleton potential θ_a in Equation (5.11) is defined for the correspondence $(\mathbf{p}_i, \mathbf{p}'_j)$ between two interesting points $\mathbf{p}_i$ and $\mathbf{p}'_j$, using their point contexts. Firstly, the affinity vectors between the corresponding elements are computed based on their distance and orientation vectors:

$$\mathcal{A}_{\mathcal{D},k}(\mathbf{p}_i, \mathbf{p}'_j) = \exp(-\frac{(\mathcal{D}_{\mathbf{p}_i}(k) - \mathcal{D}_{\mathbf{p}'_j}(k))^2}{(\max(\mathcal{D}_{\mathbf{p}_i})\sigma)^2}) \quad . \tag{5.12}$$

$$\mathcal{A}_{\Theta,k}(\mathbf{p}_i, \mathbf{p}'_j) = \exp(-\frac{(\Theta_{\mathbf{p}_i}(k) - \Theta_{\mathbf{p}'_j}(k))^2}{\delta^2}) \quad . \tag{5.13}$$

where k represents the dimension index of M-dimensional vectors $\mathcal{D}_{\mathbf{p}_i}$ and $\Theta_{\mathbf{p}_i}$ (or $\mathcal{D}_{\mathbf{p}'_j}$ and $\Theta_{\mathbf{p}'_j}$) where each dimension in $\mathcal{D}_{\mathbf{p}_i}$ (or $\mathcal{D}_{\mathbf{p}'_j}$) and $\Theta_{\mathbf{p}_i}$ (or $\Theta_{\mathbf{p}'_j}$) represents the distance and orientation of $\mathbf{p}_i$ (or $\mathbf{p}'_j$) to the k-th sample point, respectively. σ and δ are parameters to control the tolerance of distance and orientation differences, respectively. $\sigma = 0.2$ and $\delta = \pi/4$ are set in all experiments. The $\mathcal{A}_{\mathcal{D},k}$ and $\mathcal{A}_{\Theta,k}$ are calculated for M sample points and get two M-dimensional vectors $\mathcal{A}_{\mathcal{D}}(\mathbf{p}_i, \mathbf{p}'_j)$ and $\mathcal{A}_{\Theta}(\mathbf{p}_i, \mathbf{p}'_j)$. To make the value of $\mathcal{A}_{\mathcal{D}}(\mathbf{p}_i, \mathbf{p}'_j)$ invariant to scale changes, each distance difference is divided by the maximal distance in the first distance vector.

Since both $\mathcal{A}_{\mathcal{D}}$ and $\mathcal{A}_{\Theta}$ are normalised, they can be simply added to obtain the affinity vector:

$$\mathcal{A}(\mathbf{p}_i, \mathbf{p}'_j) = \mathcal{A}_{\mathcal{D}}(\mathbf{p}_i, \mathbf{p}'_j) + \mathcal{A}_{\Theta}(\mathbf{p}_i, \mathbf{p}'_j) \quad . \tag{5.14}$$

The overall similarity between $\mathbf{p}_i$ and $\mathbf{p}'_j$ can be calculated as the mean value of $\mathcal{A}(\mathbf{p}_i, \mathbf{p}'_j)$. Consequently, the singleton potential for the correspondence $(\mathbf{p}_i, \mathbf{p}'_j)$ is defined as

$$\theta_a = \theta_{\mathbf{p}_i, \mathbf{p}'_j} = \frac{1}{M} \sum_{k=1}^{M} \mathcal{A}(k) \quad . \tag{5.15}$$

The Third-order Potential: The third-order potential in Equation (5.11) is defined using angles which are formed by a triplet of interesting points. Suppose that $\mathbf{P}_1$ and $\mathbf{P}_2$ are the set of interesting points for two shapes D_1 and D_2, respectively. For any two triplets, $(\mathbf{p}_{1,i}, \mathbf{p}_{1,j}, \mathbf{p}_{1,k}) \in \mathbf{P}_1$ and $(\mathbf{p}_{2,i}, \mathbf{p}_{2,j}, \mathbf{p}_{2,k}) \in \mathbf{P}_2$, the third-order potential for each possible triple matching $(\mathbf{p}_{1,i}, \mathbf{p}_{1,j}, \mathbf{p}_{1,k}) \rightarrow (\mathbf{p}_{2,i}, \mathbf{p}_{2,j}, \mathbf{p}_{2,k})$ is defined with a truncated Gaussian kernel:

$$\theta_{abc} = \theta_{\mathbf{p}_{1,i},\mathbf{p}_{1,j},\mathbf{p}_{1,k},\mathbf{p}_{2,i},\mathbf{p}_{2,j},\mathbf{p}_{2,k}} = \begin{cases} \exp(-\gamma \| f_{i_1,j_1,k_1} - f_{i_2,j_2,k_2} \|^2) & \text{if } \| f_{i_1,j_1,k_1} - f_{i_2,j_2,k_2} \| \leqslant \vartheta \\ 0 & \text{otherwise} \end{cases} . \tag{5.16}$$

where f_{i_1,j_1,k_1} (or f_{i_2,j_2,k_2}) is the three-dimensional vector which describes sine values of three angles formed by $(\mathbf{p}_{1,i}, \mathbf{p}_{1,j}, \mathbf{p}_{1,k})$ (or $(\mathbf{p}_{2,i}, \mathbf{p}_{2,j}, \mathbf{p}_{2,k})$). Points in such a triplet are ordered in a clockwise fashion where $\mathbf{p}_{1,i}$ or $\mathbf{p}_{2,i}$ are starting points. The truncated Gaussian kernel is used to scatter and reduce matching times since the number of possible triple matching is huge and it is not necessary to compute them completely. γ is set to 2 in the experiments in Section 6.3.1. With Equation (5.16), for each triplet in $\mathbf{P}_1$, the triplets in $\mathbf{P}_2$ can be found in a neighbourhood of size ϑ.

Based on [Duc+11] and the preliminary experiments, only 20 triangles are sampled per interesting points in $\mathbf{P}_1$. There are several possible strategies to select triangles depending on user intentions. If the aim is matching with deformation allowance, the triangle should be selected at small scales. On the other hand, if one wants to capture the global property of a shape, the triangles should be big enough. According to this, the triangle should be selected based on the distribution of interesting points. If points are densely located in some regions (like the bone in Figure 4.4 (b)), more triangles are sampled in those regions. Otherwise, triangles are sampled randomly. Then, with the same strategy, the triangles of $\mathbf{P}_2$ are selected. In order to efficiently store the selected triangles, a kd-tree is employed.

5.3.1.3 Formulation Dual-Decomposition

In Equation (5.11), interesting point matching is formulated as a high-order graph matching problem combining both extrinsic similarity and intrinsic embedding information (interesting point triangles). The matching is achieved by globally optimising Equation (5.11) which includes the cost of the deformation as well as the cost of correspondences according to multiple cues. In order to obtain a globally optimal or near optimal solution while reducing the complexity without searching for all possible matching correspondences, Equation (5.11) is re-formulated into sub-problems and the high-order terms in Equation (5.11) are reduced to quadratic terms.

Specifically, the dual-decomposition method [KP09] is firstly employed to re-formulate Equation (5.11) into sub-problems that are easier to solve. A sub-problem is defined for each type of potentials, that is, $\mathcal{E}^1(\mathbf{u}|\theta^1) = \sum \theta_a x_a$ $(\theta^1 = \theta_a)$ and $\mathcal{E}^2(\mathbf{u}|\theta^2) = \sum \theta_{abc} x_a x_b x_c$ $(\theta^2 = \theta_{abc})$. Based on this definition, let v denote a problem of all sub-problems I. Under this setting, the original problem in Equation (5.11) can be approximated as the following linear combination of sub-problems:

$$\mathcal{E}(\mathbf{u}|\theta) = \sum_{v \in I} \rho_v \mathcal{E}^v(x|\theta^v) \quad . \tag{5.17}$$

where ρ_v is the weight for each sub-problem and used to control its importance. In this case, using the heuristic method of Gradient Hill Climbing integrated with Simulated Annealing [KGV83], $\rho_1 = 0.7$ and $\rho_2 = 0.3$ are set for the first and third-order sub-problems, respectively. This means that a higher priority is assigned on matching of overall shape characteristics based on the singleton potentials. Then, the original problem is solved by iteratively updating potentials θ^v and their interrelated correspondences (x_a or $x_a x_b x_c$) of each sub-problem v while fixing potentials and correspondences for the other sub-problems.

In order to solve the first-order sub-problem, the Hungarian algorithm [Kuh55] is employed. For each interesting point $\mathbf{p}_i$ in $\mathbf{P}_1$, the Hungarian algorithm can find its corresponding interesting point $\mathbf{p}'_j$ in $\mathbf{P}_2$ based on their similarity value in Equation (5.15). For the third-order sub-problem, the high-order reduction method [Ish09] is firstly employed to reduce the high-order terms in Equation (5.11) to quadratic terms. Then, the original problem in Equation (5.11) can be solved by QPBO algorithm [KR07].

Essentially, Equation (5.11) is a standard optimisation problem which could also be solved by many existing methods like HGM [ZS08], RRWHM [LCL11]. Moreover, Monte-Carlo based methods [SCL12] can also be employed to search the optimal correspondences, since there is a limited number of interesting points in each shape. Similar to the employed dual-decomposition method, these methods are designed for high-order graphs. In Section 6.3.1, the matching performance between the proposed method and other high-order matching approaches are compared. In Section 8.2, some extensions are discussed as the future work.

5.3.2 Hierarchical Skeleton Matching with Singleton and Pairwise Potentials

Benefiting from the properties of a hierarchical skeleton in Section 4.3, a shape matching method should allow more deformations on finer levels while preserving important global geometrical and topological properties. In order to do so, the proposed matching algorithm considers both global shape properties and fine-grained deformations by defining singleton and pairwise potentials for similarity computation between hierarchical skeletons. Therefore, the formulation of hierarchical skeleton matching algorithm is illustrated in the first part. Then, two potential functions are introduced which are used to capture the shape similarity from different properties of hierarchical skeletons.

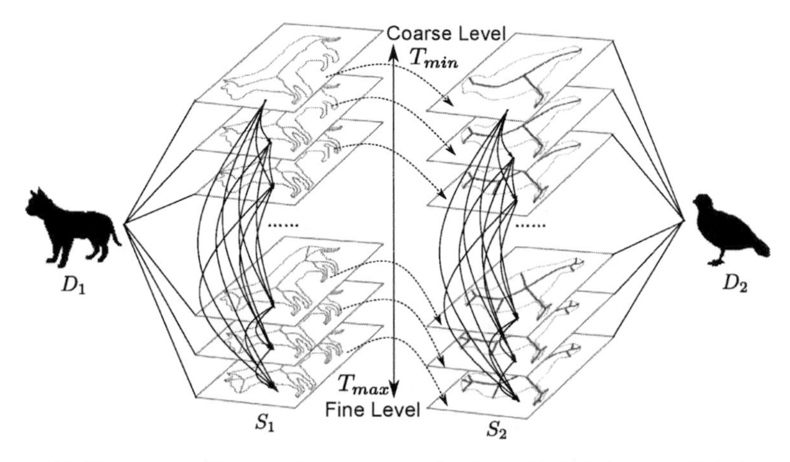

Figure 5.4: Illustration of the matching algorithm for hierarchical skeletons. All skeletons in S_1 and S_2 are ordered from coarse level (T_{min}) to fine level (T_{max}).

5.3.2.1 Formulation of Hierarchical Skeleton Matching

Let D_1 and D_2 be two planar shapes, $S(D_1)$ and $S(D_2)$ denote the full hierarchical skeletons for D_1 and D_2, respectively. Let $S_1 \subseteq S(D_1)$ or $S_2 \subseteq S(D_2)$ be the set of skeletons that are chosen from the levels $[T_{min}, T_{max}]$ (see Section 4.3) for object matching. In order to calculate the distance between S_1 and S_2, similar to the matching strategy in Section 5.3.1.1, $\mathbf{P}$ is defined as a set of correspondences between skeletons in S_1 and those in S_2. Based on this, the hierarchical skeleton matching is formulated as follows:

$$d(S_1, S_2) = \sum_{a \in \mathbf{P}} g(\theta_a)x'_a + \sum_{(a,b) \in \mathbf{P} \times \mathbf{P}} g(\theta_{ab})x'_a x'_b \quad . \tag{5.18}$$

where θ_a is the matching cost for each correspondence $a \in \mathbf{P}$ (the singleton potential, dotted arrows in Figure 5.4) that expresses the property of skeleton-based object matching. θ_{ab} is the matching cost of a pair of correspondences $(a, b) \in \mathbf{P} \times \mathbf{P}$ (the pairwise potential, skeleton pairs connected by solid arrows in Figure 5.4) that represents the difference between skeleton evolutions in two hierarchical skeletons. Here, the skeleton evolution of a hierarchical skeleton is characterised as the changes between skeleton pairs on two different levels. The rationale behind this is that there is no predictable pattern in how skeletons change in different levels. In some levels, skeletons are even the same due to the overlapped removal of skeleton points along with the polygon simplification. However, as shown in Figure 4.15, the overall trend is that skeletons are gradually becoming more complex along with the DCE steps. Built on this observation, the skeleton changes are collected by considering all skeleton pairs in a hierarchical skeleton. With this strategy, even if the skeleton changes are zero within some pairs, the skeleton evolution information is still sufficient to distinguish two hierarchical skeletons since the overall accumulation of the skeleton changes are collected. $g(\theta_a) = |\theta_a|^\alpha$ (or $g(\theta_{ab}) = |\theta_{ab}|^\alpha$) is the power function term to alleviate effects by abnormally large matching costs θ_a (or θ_{ab}) [Jeg+12]. Based on the preliminary experiment, α is set to 0.18. This value is obtained by employing the optimisation method introduced in Section 5.2.2.3. x'_a is the boolean indicator variable:

$$x'_a = \begin{cases} 1 & \text{if } a = (i,j) \in \mathbf{P} \text{ and } i = j \\ 0 & \text{otherwise} \end{cases} . \tag{5.19}$$

where i and j denote the skeleton level, $i, j \in [T_{min}, T_{max}]$. In the following discussion, the skeleton on the ith level of S_1 and on the jth level of S_2 are denoted by $S_{1,i}$ and $S_{2,j}$, respectively.

As shown in Equation (5.19), $x'_a = 1$ is set for the constraint condition $i = j$. This constraint ensures that each skeleton in S_1 is mapped to the skeleton on the same level in S_2 (presented by dash arrows in Figure 5.4). The rationale behind this is as follows: Firstly, it is very important for the proposed method to reduce the computation complexity since one computation of the similarity between two skeletons takes a long time, so repeating this computation $(T_{max} - T_{min} + 1)^2$ times requires prohibitive computational costs. Secondly, a hierarchical skeleton is organised from simple to complex, and the skeleton on one level is included in the ones on higher levels. In other words, even if one hierarchical level does not offer the best correspondences for endpoints in two skeletons, they could be found on the higher levels. Thirdly, experimental results in Figure 6.14 illustrate that the performance of finding the global optimum matching operates far less efficiently than the proposed method due to the overfitting problem. Another possible singleton potential is to match skeletons on the levels that are adjacent to the current level. However, as illustrated in Figure 4.15, many adjacent skeletons have the same skeleton structure. Moreover, the newly added endpoints in the adjacent skeletons could be jumped by the skeleton matching algorithm which is employed for the singleton potential. Thus, matching skeletons on the adjacent levels will not offer significant performance improvement, but it incurs a significant increase of computational cost. Therefore, matching skeletons on the same level is reasonable in terms of both the computational cost and performance.

5.3.2.2 Potential Functions

The Singleton Potential: For each correspondence (i, j), $i, j \in [T_{min}, T_{max}]$, the skeleton graph information is considered to define its singleton potential as in [BL08]. The idea is to find the best matching between endpoints in two skeletons. The skeleton graphs $S_{1,i}$ and $S_{2,j}$ are matched by comparing the geodesic paths between their skeleton endpoints. Then, all the dissimilarity values between their endpoints are represented as a distance matrix $\mathcal{U}(S_{1,i}, S_{2,j})$. The total dissimilarity $\mathcal{D}(S_{1,i}, S_{2,j})$ between $S_{1,i}$ and $S_{2,j}$ is computed by searching correspondences between skeleton endpoints with the Hungarian algorithm [Kuh55] on $\mathcal{U}(S_{1,i}, S_{2,j})$, so that endpoints in $S_{1,i}$ and $S_{2,j}$ are matched with the minimal cost. The singleton potential for the correspondence (i, j) is defined as

$$\theta_a = \sigma \cdot \mathcal{D}(S_{1,i}, S_{2,j}) \quad . \tag{5.20}$$

where σ is a weighting factor obtained using the arithmetic progression $[\frac{1}{T_{max}-T_{min}+1}, \cdots, 1]$. In other words, the higher weights are assigned to coarse-level skeletons to preserve important global shape properties while charging lower weights on fine-level skeletons to allow small local deformations. Although the arithmetic progression uses the same weight for skeletons on a certain level without considering their characteristics, it practically works well. A possible improvement is to estimate optimal weights depending on a given pair of hierarchical skeletons using distance metric learning [Xin+03]. In Section 8.2, this extension is discussed as a future work.

The Pairwise Potential: The pairwise potential is calculated by comparing skeletons ($S_{1,i}$ and $S_{1,j}$) (solid arrows on S_1 in Figure 5.4) on two levels in S_1 to those ($S_{2,i}$ and $S_{2,j}$) (solid arrows on S_2 in Figure 5.4) on the respective levels in S_2:

$$\theta_{ab} = \tfrac{1}{2} o_{i,j} \quad . \tag{5.21}$$

where $i, j \in [T_{min}, T_{max}]$ and $o_{i,j}$ denotes the similarity between the pair of $S_{1,i}$ and $S_{1,j}$ and the pair of $S_{2,i}$ and $S_{2,j}$:

$$o_{i,j} = \frac{|f(S_{1,i}, S_{1,j}) - f(S_{2,i}, S_{2,j})|}{f(S_{1,i}, S_{1,j}) + f(S_{2,i}, S_{2,j})} \quad . \tag{5.22}$$

In Equation (5.22), $f(S_{z,i}, S_{z,j}) (z \in \{1, 2\})$ represents the change from $S_{z,i}$ to $S_{z,j}$. $f(S_{z,i}, S_{z,j})$ is calculated based on the radius and length of new skeleton points between $S_{z,i}$ and $S_{z,j}$ (Figure 5.5 (a) and (b)). Let N_O denote the number of skeleton points of the new skeleton branches (e.g. the blue line in Figure 5.5 (b)) from $S_{z,i}$ to $S_{z,j}$. The dissimilarity $f(S_{z,i}, S_{z,j})$ is defined as

$$f(S_{z,i}, S_{z,j}) = \sum_{e=1}^{N_0} r_e + \eta^{\frac{(l(S_{z,i})-l(S_{z,j}))^2}{l(S_{z,i})+l(S_{z,j})}} \quad . \tag{5.23}$$

where $l(S_{z,i})$ and $l(S_{z,j})$ denote the length of skeleton $S_{z,i}$ and that of $S_{z,j}$, respectively. In order to ensure that the proposed method is invariant to scale changes, the new branch lengths are normalised. r_e denotes the radius (e.g. the dotted arrow in Figure 5.5 (b)) of the tangent disk of a skeleton point that touches the shape boundary in two or more locations. η is the weight factor. This parameter is used for controlling the numeric consistency between the term of radii and the term of skeleton lengths. Without η, the sum of radii will be much larger than the distance between two skeletons and finally dominate $f(S_{z,i}, S_{z,j})$ in Equation (5.23). Based on the preliminary experiments in [Yan+16c], η is set to 1.2. In Equation (5.23), the difference between skeleton lengths primarily captures their coarse-grained dissimilarity, while their fine-grained difference is magnified by taking the sum of radii from skeleton points. By fusing two terms, the distinctiveness between two skeletons is more obvious and robust.

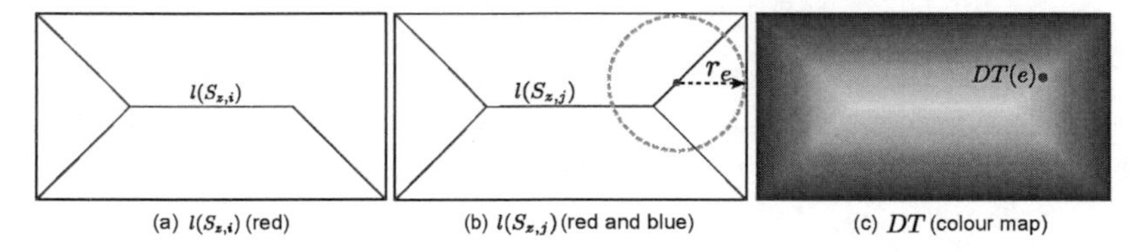

Figure 5.5: Radius and skeleton length. Skeleton $S_{z,j}$ has one new branch (blue line) compared to skeleton $S_{z,i}$. $l(S_{z,i})$ is calculated by the length of red lines in (a) while $l(S_{z,j})$ is calculated by the length of the red lines plus the length of the blue line in (b). For a skeleton point (red point), r_e is the radius (dotted arrow) of its tangent disk (green dotted circle). r_e is approximately equal to the normalised value of $DT(e)$ in the distance transform matrix DT.

It should be noted that the length ratio pairwise potential have likewise been used extensively [DPS00; TH04], however, the sequence of radii is not fully involved. The radius is considered to better represent the reconstruction area between skeletons on different hierarchical levels. For this purpose, the distance transform is firstly performed to compute a matrix DT in which each pixel in the original shape D is characterised by the distance to its closest boundary (Figure 5.5 (c)). Then, for each skeleton point with index e, $DT(e)$ is selected as the radius r_e. To make the proposed method invariant to the scale, a r_e is normalised by the following way:

$$r_e = \frac{DT(e)}{\frac{1}{N''} \sum_{v=1}^{N''} DT(q_v)} \quad . \tag{5.24}$$

where $q_v(v = 1,2,\cdots,N'')$ varies over all N'' pixels in D.

Although flexible matchings for measuring skeleton evolution between S_1 and S_2 are possible using tree or graph matching algorithms [BL08; LG99], they are not employed for the following reasons: First, tree or graph matching algorithms can flexibly match $S_{1,i} \in S_1$ and $S_{2,j} \in S_2$ where i and j do not have to be the same. However, this kind of matching requires examining similarities for many pairs of $S_{1,i}$ and $S_{2,j}$, and what is worse, even computing the similarity for one pair constitutes a high computational cost. Secondly, tree or graph matching methods normally consider the correspondences between endpoints in $S_{1,i}$ and $S_{2,j}$. However, as can be seen in Figure 4.15, a skeleton on one level is characterised by adding a trivial endpoint to the skeleton on a higher level. Such an endpoint is not useful for describing the transitions of skeletons (i.e. skeleton evolution). Compared to endpoints, changes of length and radius are considered to be a better representation of skeleton evolution, and they cannot be used directly in usual tree or graph matching algorithms.

5.4 Shape Matching using Integrated Features

In this section, three integrated descriptors are introduced for object retrieval based on the proposed and some existing shape descriptors. Integrated descriptors normally have a higher discriminating power than the individual one since more shape features are captured and preserved in the integrated descriptors. However, not every descriptor can be integrated for the following two reasons: (1) Descriptor integration should not incur too much computational complexity. In this case, if two shape descriptors both need high time complexity for feature generation, they are not suited to be integrated. (2) Descriptor integration should not trigger the overfitting problem. For this, a fusing weight is normally required to control the impact power of each fused descriptor. Based on the above analysis, in this section, the coarse- and fine-grained shape descriptors are selected for integrating. Due to the difference between their feature structures, the original matching algorithms are employed for calculating similarity (or dissimilarity) between two shapes based on their correlated shape descriptors. After that, the calculated similarities (or dissimilarities) are fused together as the final similarity (or dissimilarity) of the fused shape descriptors. Experiments in Section 6.4 illustrate the performance improvements of the proposed three integrated descriptors.

5.4.1 Shape Matching with Shape Context and Boundary Segments

As illustrated in Figure 5.6, the general ideal of this integrated descriptor is based on the shape contour interesting points (Section 3.2), shape context feature [BMP02] and the open curve descriptor (Section 3.2). Specifically, based on the contour interesting points (Figure 5.6 (a))

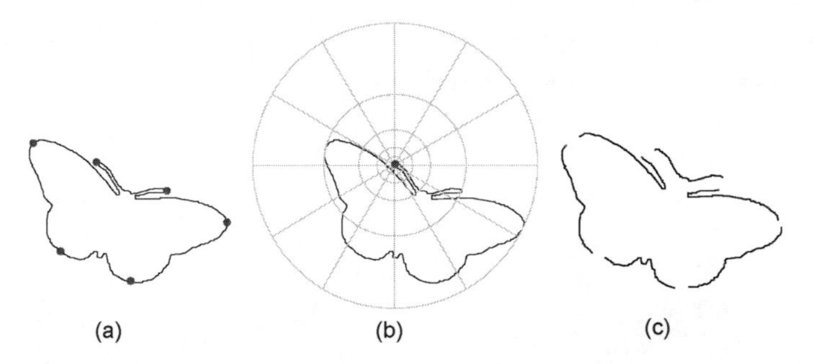

Figure 5.6: Idea of integrating shape context and boundary segment descriptors.

introduced in Section 3.2, a shape boundary is partitioned into several boundary segments (Figure 5.6 (c)). In order to describe each interesting point, the shape context descriptor is employed (Figure 5.6 (b)). Moreover, the proposed open curve descriptor in 3.2 is employed to represent each boundary segment.

For shape matching, the interesting points and boundary segments are matched independently using the shape context matching algorithm in [BMP02] and the proposed open curve matching method in Section 5.2.1, respectively. After matching, the global shape distance is estimated as the weighted sum of two terms: interesting points' dissimilarity and boundary segments' dissimilarity. Particularly, the distance between two shapes $D^\star$ and $D^\diamond$ with a fused scheme between their interesting points and boundary segments is defined as:

$$d(D^\star, D^\diamond) = \eta d_{points}(D^\star, D^\diamond) + (1-\eta) d_{segments}(D^\star, D^\diamond) \quad . \tag{5.25}$$

where d_{points} and $d_{segments}$ denote the dissimilarity between interesting points and boundary segments, respectively. η is a weight for fusing which is assigned by the optimisation method introduced in Section 5.2.2.3. By the preliminary experiments in [Fei+14], η is set to 0.7 and is employed for all experiments in Section 6.4.1. For the task of shape retrieval, given a set of M shapes, the shape distance method in Equation (5.25) is applied to obtain a $M \times M$ distance matrix to describe the pairwise relations between all shapes in terms of a dissimilarity measure. Such an approach ignores the fact that also distances to all other shapes contain important information about the overall shape manifold. Therefore, in Section 6.4.1, the Mutual kNN Graph (MG) method [KDB10] is employed to improve the shape retrieval performance by analysing the underlying structure of the shape manifold. This method captures the manifold structure of the data by defining a neighbourhood for each data point in terms of a modified version of a mutual kNN graph which yields an improved performance on all databases in Section 6.4.1.

5.4.2 Shape Matching with Skeleton and Boundary Segments

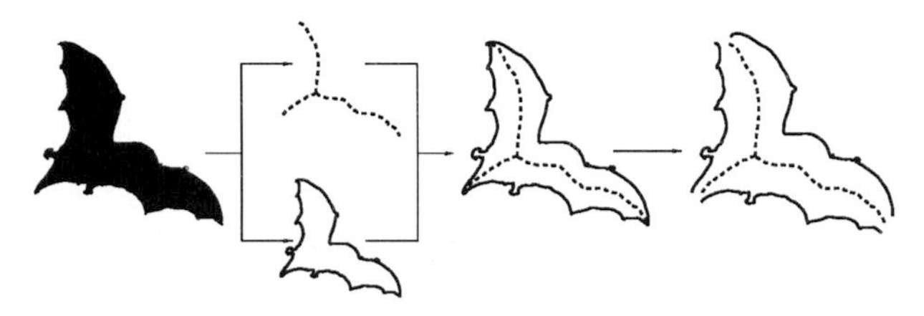

Figure 5.7: Idea of integrating shape skeleton and boundary segment descriptors.

The main idea of this descriptor is to use skeleton endpoints for shape boundary partition (Figure 5.7). Then a shape is represented by a skeleton graph and boundary segments which can be described by the open curve descriptor in Section 3.2. As introduced in Section 4.3, a shape skeleton normally requires a pruning process to remove redundant skeleton branches. This process is also required in this integrated descriptor since without skeleton pruning, the partitioned boundary segments are too small to be discriminated. Thus, the DCE-based skeleton pruning method is employed. However, this method requires a stop parameter k to control the pruning power and it is impractical to assign a value to each shape. Based on the experiments in Section 6.3.2.1, $k = 10$ is employed for generating the object skeleton in this part since it achieves the best performance on the skeleton graph matching in Table 6.27. Finally, the matching of two shapes $D^\star$ and $D^\diamond$ is performed separately for their skeleton and boundary segment representations.

For skeleton graph matching, the Path Similarity Skeleton Graph Matching (PSSGM) [BL08] method is employed for calculating dissimilarity $d_{skeleton}$ between two shapes using their skeletons. For boundary segment matching, the proposed open curve representation and matching methods in Section 3.2 and Section 5.2.1 are employed for calculating dissimilarity $d_{segments}$ between two shapes using their boundary segments. Similar to Equation (5.25), the final dissimilarity between $D^\star$ and $D^\diamond$ is calculated by

$$d(D^\star, D^\diamond) = \eta d_{skeleton}(D^\star, D^\diamond) + (1 - \eta) d_{segments}(D^\star, D^\diamond) \quad . \tag{5.26}$$

Based on the preliminary experiments in [Yan+14b], η is set to 0.5 for all experiments in Section 6.4.2.

5.4.3 Shape Matching with Shape Context and Bounding Boxes

In this part, the SC descriptor [BMP02] and the proposed bounding boxes-based descriptor in Section 3.3 are fused together for shape representation and matching. Specifically, for the SC descriptor, a shape is represented by a set of points sampled from the shape contour. Each sample point is represented by the shape context to describe the geometrical and topological properties of its location. Finding correspondences between two shapes is then equivalent to finding for each sample point on one shape the sample point on the other shape that has the most similar shape context. Based on the correspondences, the dissimilarity d_{SC} between two shapes $D^\star$ and $D^\diamond$ is computed as the sum of matching errors between corresponding points, together with a term measuring the magnitude of the align transform from one shape to another. For the bounding boxes-based descriptor in Section 3.3, the proposed matching algorithm in Section 5.2.2.1 is employed to calculate the dissimilarity d_{boxes} between $D^\star$ and $D^\diamond$. Similar to Equation (5.25) and Equation (5.26), the final dissimilarity is computed by:

$$d(D^\star, D^\diamond) = \eta d_{SC}(D^\star, D^\diamond) + (1-\eta) d_{boxes}(D^\star, D^\diamond) \quad . \tag{5.27}$$

where $\eta = 0.7$ [Yan+15b] is set for all experiments in Section 6.4.3.

5.5 Summary

In this chapter, several shape matching methods are proposed based on the proposed shape descriptors in Chapter 3 and 4. For coarse-grained shape descriptors, since those descriptors normally have simple feature structures, shape matching methods are applied based on vector distance or certain statistical methods. Therefore, shape matching with coarse-grained descriptors has low computational complexity. Compared to coarse-grained shape descriptors, fine-grained descriptors normally have a complex feature structure. Moreover, some hidden properties of fine-grained descriptors should be explored to improve the shape matching performance. Therefore, matching algorithms for fine-grained shape descriptors should be specially designed built on their feature properties.

In Section 5.2, compared to traditional shape matching methods for coarse-grained descriptors, the proposed matching methods have higher adaptability on different datasets. This is benefiting from the parameters that control the discriminating power of each feature in a feature vector. In order to find proper values for the parameters, a supervised optimisation strategy is proposed. Experiments in Section 6.2 validate the capability of this strategy.

In order to fully use the properties of interesting point-based and hierarchical skeleton-based shape descriptors in Chapter 4, two meaningful shape matching methods are proposed in Section 5.3. For the interesting point-based descriptors, a high-order graph matching method is employed in order to involve not only the assignments of single interesting points, but also the geometrical relations of their triplets. For the hierarchical skeleton-based descriptor, a matching method is developed which considers similarities for both single skeletons and skeleton pairs (skeleton evolution) in a hierarchical skeleton. Experiments in Section 6.3 on several datasets demonstrate that the proposed methods are significantly superior to most conventional fine-grained shape descriptors and their matching methods. In the last section of this chapter, shape matching using three fused descriptors are introduced. Those descriptors are normally fused by coarse- and fine-grained descriptors to enhance the discriminating power above the original one without adding too much computational complexity. Experiments in Section 5.4 prove the efficiency of those methods.

Chapter 6

Experiments and Results

This part collects the experiments and results which are mentioned during the shape generation, representation and matching process. The detailed experimental setups as well as parameters are also introduced in this chapter. Based on the results, further analyses and discussions are addressed to prove strengths and weaknesses of the proposed methods.

6.1 Shape Contour Detection by Open Curves Matching

There are two experiments involved during the shape generation process in Chapter 2. In the first one, CS descriptors are evaluated for open curve matching. Based on the evaluation results, a competent CS descriptor and its matching algorithm are used for open curve matching as well as shape contour detection. The detected results are reported in the second experiment.

6.1.1 Evaluating Contour Segment Descriptors

This section presents the experimental environments and results in the experimental settings defined in Section 2.3.3.

6.1.1.1 Datasets and Environment

To the best of my knowledge, there are no suitable datasets for evaluating the performance of CS descriptors. This is because in most existing CS-related applications [CFT08; Wan+12b; Lu+09b; Zhu+08; PT10; RDB10b; PKB11], researchers propose CS descriptors only for their own specific scenarios. Moreover, the existing datasets [DRB10b; ML11b; MSJB15; Yan+14b] cannot be directly employed for CS evaluation since they contain only images or shapes rather than CS. Thus, there are two datasets specially designed for CS and open curve matching experiments.

MPEG7 CS: This dataset is designed for evaluating CS matching. The MPEG7 [LLE00] dataset is a standard and commonly used shape dataset for evaluating shape matching and classification. The total number of images in the MPEG7 database is 1400: 70 classes of various shapes, each class with 20 images. As shown in Figure 6.1, the MPEG7 CS dataset is created using shape contours. For the shapes from the same class in MPEG7, their contours are firstly extracted and then the same part of contours are manually removed. Finally, the MPEG7 CS dataset is generated with 1400 CSs. For easy and fair performance evaluation of CS descriptors, the generated CSs have the same number of CS points. With this property, this dataset is also used for comparing the runtime between CS descriptors.

Figure 6.1: Sample shapes from the MPEG7 database [LLE00] for generating CSs. The correlated shape contours and CSs are shown in the second and third row.

The CS matching performances are evaluated by applying CS retrieval in this dataset. In order to quantitatively evaluate the matching performance, the so-called bulls-eye score [LLE00] is employed for evaluation. Specifically, given a query CS, the 40 most similar CSs are retrieved from the database and the number of CSs belonging to the same class as the query are counted. The bulls-eye score is the ratio of the total number of correctly matched CSs to the number of all the possible matches (which is 20×1400). Thus, the best score is 100 percent. However, as discussed in [LLE00], the 100% bulls-eye score is impossible to obtain since some classes contain objects whose shapes are significantly different so that it is not possible to group them into the same class using only their shapes.

ETHZ CS: This dataset is designed for evaluating open curve matching. ETHZ [FTG06] is a dataset for testing object class detection algorithms. It contains 255 test images and features five diverse shape-based classes (apple logos, bottles, giraffes, mugs and swans). Based on the shapes from ground truth, the following open curves are manually generated by keeping their contour parts in the parentheses (Figure 6.2): 44 open curves for apple logos (the right half part), 55 open curves for bottles (the right half part), 91 open curves for giraffes (the upper neck part), 48 open curves for mugs (the right half part with the handle) and 32 open curves for swans (the upper half part with the head). In total, there are 271 open curves.

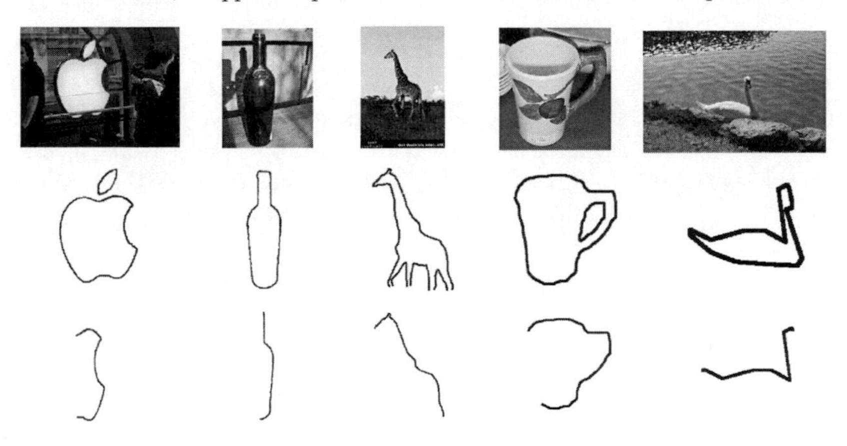

Figure 6.2: Sample images from the ETHZ database [FTG06]. The correlated object contours and CSs are shown in the second and third row.

The experiment on this dataset aims to evaluate the open curve matching performances using different CS descriptors. To do so, an open curve retrieval scenario is conducted using the open curve matching method introduced in Section 2.3. Specifically, each of the 271 curves is used as a query and the 100 most similar curves among the whole dataset are retrieved. The final evaluation is composed of two parts: recall and precision. Both parts are calculated by the mean value of recall and precision from all classes [Faw06]. For example, since there are 44 open curves in the Apple logo class, the precision (or recall) for this class is calculated by taking the average of precisions (or recalls) that are obtained using each of the 44 open curves as a query.

The experiments in this section are delivered on two platforms: Cluster and laptop. Feature generation and full object retrieval experiments are accomplished on *Horus*, a cluster provided by the University of Siegen, which includes 136 nodes, each consisting of 2 Intel Xeon X5650 with 2,66 GHz and 48 GB memory. With this cluster, the massive experiments can be finished efficiently. In order to fairly compare the runtime of each CS descriptor, the runtime estimation experiments are finished on a laptop with Intel Core i7 2.2GHz

CPU, 8.00GB memory and 64-bit Windows 8.1 OS. Here the cluster is not used for runtime estimation, mainly because the task distribution in *Horus* may cause an unfair comparison since various processes in each node are different. All methods in the experiment are implemented in Matlab. In the *Horus*, codes are executed in Matlab R2014b 64-bit, while Matlab R2015a 64-bit is used on the laptop.

6.1.1.2 Experiment 1: Invariant Properties

Table 6.1: Invariance properties of different CS descriptors in Section 2.3.1. SI, RI and TI denote the scale invariance, rotation invariance and translation invariance. The existence and lack of an invariance are indicated with '+' and '-', respectively. Regarding +* of $\mathbf{f}_{14}$, if the sampling is dense enough, then it is rotation invariant, and vice versa.

Simple CS Descriptors				Signature-based CS Descriptors				Rich CS Descriptors			
Descriptors	SI	RI	TI	Descriptors	SI	RI	TI	Descriptors	SI	RI	TI
$\mathbf{f}_1$	+	+	+	$\mathbf{f}_{10}$	-	-	-	$\mathbf{f}_{18}$	+	+	+
$\mathbf{f}_2$	+	+	+	$\mathbf{f}_{11}$	-	+	+	$\mathbf{f}_{19}$	+	+	+
$\mathbf{f}_3$	+	+	+	$\mathbf{f}_{12}$	+	+	+	$\mathbf{f}_{20}$	+	-	+
$\mathbf{f}_4$	+	+	+	$\mathbf{f}_{13}$	+	+	+	$\mathbf{f}_{21}$	+	+	+
$\mathbf{f}_5$	+	+	+	$\mathbf{f}_{14}$	-	+*	+	$\mathbf{f}_{22}$	+	+	+
$\mathbf{f}_6$	+	+	+	$\mathbf{f}_{15}$	-	+	+	$\mathbf{f}_{23}$	+	+	+
$\mathbf{f}_7$	+	+	+	$\mathbf{f}_{16}$	+	+	+	$\mathbf{f}_{24}$	+	+	+
$\mathbf{f}_8$	+	+	+	$\mathbf{f}_{17}$	+	+	+	$\mathbf{f}_{25}$	+	+	+
$\mathbf{f}_9$	+	+	+					$\mathbf{f}_{26}$	+	+	+

Table 6.1 illustrates the theoretical invariance properties of different CS descriptors introduced in Section 2.3.1. Firstly, all the CS descriptors are invariant to translation of CSs, except for Comcoor ($\mathbf{f}_{10}$) which is represented by coordinates of CS points ($\mathbf{f}_{10}$ is also neither scaling nor rotation invariant). Secondly, it can be clearly observed that all simple CS descriptors are invariant to rotation since they are generated by only considering the overall feature of a CS. Moreover, a normalisation process can ensure the scaling invariance of simple descriptors. Thirdly, some signature-based descriptors, such as the Area Function ($\mathbf{f}_{14}$) and Triangle Area ($\mathbf{f}_{15}$), do not perform well for scaled CSs. In practice, they can be normalised by the CS length. These descriptors thereby become scale-invariant. Lastly, except Contour Context ($\mathbf{f}_{20}$), all the rich descriptors comply with the three invariance properties. In practice, the rotation property of the Contour Context descriptor can be improved by the preprocessing method introduced in [Yan+14b].

6.1.1.3 Experiment 2: Matching Performance

Table 6.2: CS matching results (%) using simple CS descriptors on the MPEG7 CS dataset. As introduced in Table 2.1, L1-L4 denote the sampling densities that are used for generating CS descriptors.

Descriptors	L1	L2	L3	L4
$\mathbf{f}_1$	3.7	3.7	3.7	3.7
$\mathbf{f}_2$	3.7	3.7	3.7	3.7
$\mathbf{f}_3$	**22.3**	**22.8**	**22.8**	**22.8**
$\mathbf{f}_4$	**22.9**	**24.0**	**24.1**	**23.4**
$\mathbf{f}_5$	**22.3**	**22.8**	**22.8**	**22.8**
$\mathbf{f}_6$	18.7	18.9	18.9	18.9
$\mathbf{f}_7$	16.6	15.9	15.7	16.0
$\mathbf{f}_8$	3.7	3.7	3.7	3.7
$\mathbf{f}_9$	18.7	18.9	18.9	18.9

Table 6.2 illustrates the CS retrieval results on the MPEG7 CS dataset using simple CS descriptors. It can be seen that Eccentricity ($\mathbf{f}_3$), Bending ($\mathbf{f}_4$) and Rectangularity ($\mathbf{f}_5$) outperform the other descriptors in which Bending ($\mathbf{f}_4$) achieves the best performance on the MPEG7 CS dataset. Moreover, considering the performance on different lengths, it can be observed that for each simple descriptor, the scores are almost the same. Therefore, all the simple CS descriptors are robust to CS length changes. The rationale behind this is that simple CS descriptors are calculated by considering only global coarse-grained CS features. Even if some fine-grained features are lost because of a small number of sample points, the global features remain the same.

For the signature-based CS descriptors on the MPEG7 CS dataset, in Table 6.3, Comcoor ($\mathbf{f}_{10}$) with the Hungarian matching method achieves the best bulls-eye score. In Table 6.4, both Comcoor ($\mathbf{f}_{10}$) and Cendistance ($\mathbf{f}_{11}$) with the Hellinger distance method obtain the best score. Among all results, Comcoor ($\mathbf{f}_{10}$) with the Hungarian matching method achieves the best performance (64.4% bulls-eye score). Similar to the ETHZ dataset, it can be observed that the CS matching performance is enhanced and damaged dramatically if the matching algorithms or vector distance methods are changed. Moreover, for most signature-based CS descriptors, CS retrieval using point matching methods performs much better than the vector distance methods.

Table 6.5 and Table 6.6 illustrate the CS retrieval performance using rich CS descriptors. Among all rich descriptors, Beam Angle ($\mathbf{f}_{21}$) achieves promising results in all three matching methods, in which the Hungarian matching method yields the best score (79.6% bulls-eye

Table 6.3: CS matching results (%) with signature-based CS descriptors using point matching methods on the MPEG7 CS dataset.

Descriptors	DTW				DP				Hungarian			
	L1	L2	L3	L4	L1	L2	L3	L4	L1	L2	L3	L4
$\mathbf{f}_{10}$	62.1	62.3	62.3	62.4	50.0	49.9	50.6	50.9	**63.3**	**64.1**	**64.2**	**64.4**
$\mathbf{f}_{11}$	60.4	61.6	61.8	61.8	58.1	60.7	61.1	61.4	48.0	48.3	48.3	48.3
$\mathbf{f}_{12}$	54.5	55.9	60.0	61.4	54.9	55.7	59.6	60.8	51.1	51.4	50.0	49.8
$\mathbf{f}_{13}$	50.3	52.8	54.8	55.9	49.3	52.1	54.2	55.0	46.0	48.9	49.6	48.9
$\mathbf{f}_{14}$	55.0	57.7	57.7	59.4	50.8	54.5	53.7	56.4	41.9	43.6	38.6	42.4
$\mathbf{f}_{15}$	58.3	59.6	60.0	59.9	53.8	55.7	56.3	56.6	49.0	49.4	49.6	49.7
$\mathbf{f}_{16}$	3.2	2.9	2.9	2.9	3.3	3.0	2.9	2.9	3.3	2.9	2.9	2.9
$\mathbf{f}_{17}$	15.4	14.5	12.4	10.8	41.1	40.3	36.3	32.4	38.5	39.9	39.9	40.6

Table 6.4: CS matching results (%) with signature-based CS descriptors using vector distance methods on the MPEG7 CS dataset.

Descriptors	Correlation				HI				χ^2-Statistics				Hellinger			
	L1	L2	L3	L4	L1	L2	L3	L4	L1	L2	L3	L4	L1	L2	L3	L4
$\mathbf{f}_{10}$	8.2	7.1	7.1	6.9	13.5	14.0	14.5	14.3	55.7	55.8	56.5	56.6	**63.7**	**63.6**	**63.7**	**63.5**
$\mathbf{f}_{11}$	5.6	5.6	5.6	5.5	5.5	5.4	5.4	5.4	63.0	62.7	62.8	62.6	**63.7**	**63.6**	**63.7**	**63.5**
$\mathbf{f}_{12}$	5.9	5.8	5.9	5.9	7.1	7.0	6.8	6.6	44.2	47.4	50.0	49.4	42.7	41.8	44.8	43.7
$\mathbf{f}_{13}$	6.0	6.3	6.3	6.2	6.0	8.7	14.0	19.0	35.1	20.6	16.3	12.2	55.3	52.6	49.6	45.7
$\mathbf{f}_{14}$	5.6	5.6	6.2	5.6	5.1	5.1	5.1	5.1	54.4	54.8	54.8	54.3	49.7	49.2	47.7	46.3
$\mathbf{f}_{15}$	5.8	5.9	5.7	5.6	5.5	5.5	5.5	5.7	36.2	34.5	35.3	32.1	51.1	51.8	51.9	51.3
$\mathbf{f}_{16}$	3.3	3.0	2.9	2.9	2.9	2.9	2.9	2.9	3.2	2.9	2.9	2.9	2.9	2.9	2.9	2.9
$\mathbf{f}_{17}$	44.6	39.5	33.0	27.3	48.9	37.5	28.0	24.7	0.07	0.15	0.17	0.18	42.0	30.4	22.3	18.1

score). For most rich descriptors ($\mathbf{f}_{18}$ - $\mathbf{f}_{26}$), they are relatively robust to the CS length and matching algorithm changing.

Comparing the performance of three types of CS descriptors in the MPEG7 CS dataset, the following observations can be drawn: (1) Most rich descriptors have a better performance than the signature-based descriptors. (2) Most signature-based CS descriptors outperform the simple CS descriptors. (3) For most signature-based CS descriptors, the point matching strategy performs better than the vector distance strategy.

Table 6.7 shows Precisions (P) and Recalls (R) of simple CS descriptors. As seen from this figure, Eccentricity ($\mathbf{f}_3$) and Rectangularity ($\mathbf{f}_5$) outperform the other descriptors (90.1% recall). Moreover, considering the performance on different lengths, like the MPEG7 CS dataset, all simple CS descriptors are robust to CS length changing.

For the signature-based CS descriptors, as shown in Table 6.8 and Table 6.9, it can

Table 6.5: CS matching results (%) using rich CS descriptors and point matching methods on the MPEG7 CS dataset.

Descriptors	DTW				DP				Hungarian			
	L1	L2	L3	L4	L1	L2	L3	L4	L1	L2	L3	L4
$\mathbf{f}_{18}$	64.9	64.7	63.7	63.4	70.8	74.8	76.2	76.9	62.9	70.7	72.7	74.0
$\mathbf{f}_{19}$	40.9	39.0	36.9	35.8	68.5	69.5	70.3	70.5	69.8	70.1	70.4	70.3
$\mathbf{f}_{20}$	64.2	64.5	64.5	64.5	62.4	63.9	64.2	64.4	66.2	67.7	68.1	68.4
$\mathbf{f}_{21}$	72.8	75.1	74.6	73.6	73.1	75.1	75.3	73.8	**79.6**	**79.6**	**77.4**	**73.9**
$\mathbf{f}_{22}$	64.6	64.6	64.8	64.4	69.2	73.3	76.6	77.4	64.8	66.4	67.2	66.5
$\mathbf{f}_{23}$	57.6	60.1	61.0	61.2	48.8	52.6	54.2	54.2	47.4	51.5	53.4	53.5
$\mathbf{f}_{24}$	66.3	66.0	65.9	65.3	70.6	73.8	76.5	77.3	69.4	72.8	74.3	75.0
$\mathbf{f}_{25}$	64.2	64.6	64.7	64.8	64.1	64.3	64.3	64.4	64.1	64.3	64.4	64.4

Table 6.6: CS matching results (%) using rich descriptor and vector distance methods on the MPEG7 CS dataset.

Descriptors	Correlation				HI				χ^2-Statistics				Hellinger			
	L1	L2	L3	L4	L1	L2	L3	L4	L1	L2	L3	L4	L1	L2	L3	L4
$\mathbf{f}_{26}$	70.8	70.7	70.5	70.3	70.0	71.0	71.2	71.2	72.7	73.2	73.3	73.3	74.5	74.8	74.8	74.8

Table 6.7: Open curve matching results (%) using simple CS descriptors on the ETHZ CS dataset.

Descriptors	L1		L2		L3		L4	
	P	R	P	R	P	R	P	R
$\mathbf{f}_1$	33.9	56.8	34.0	56.9	34.1	57.1	34.0	57.0
$\mathbf{f}_2$	42.5	75.1	42.8	75.5	42.7	75.5	42.7	75.4
$\mathbf{f}_3$	**55.0**	**90.0**	**55.1**	**90.1**	**55.1**	**90.1**	**55.1**	**90.1**
$\mathbf{f}_4$	48.1	80.7	48.8	81.7	48.9	81.9	49.3	82.4
$\mathbf{f}_5$	**55.0**	**90.0**	**55.1**	**90.1**	**55.1**	**90.1**	**55.1**	**90.1**
$\mathbf{f}_6$	52.5	87.9	52.5	87.8	52.5	87.9	52.5	87.9
$\mathbf{f}_7$	37.1	61.1	37.1	60.8	36.1	58.8	36.4	59.7
$\mathbf{f}_8$	38.1	65.7	38.1	65.6	38.2	65.8	38.2	65.8
$\mathbf{f}_9$	52.5	87.9	52.5	87.8	52.5	87.9	52.5	87.9

be clearly observed that for both point matching and vector distance strategies, matching performances are highly related to the matching methods. For two strategies, point matching methods perform better than the vector distance methods. Specifically, Tangent ($\mathbf{f}_{12}$) with dynamic programming and Comcoor ($\mathbf{f}_{10}$) with χ^2-Statistics achieve the best performance

Table 6.8: Open curve matching results (%) with signature-based CS descriptors using point matching on the ETHZ CS dataset.

Descriptors		DTW				DP				Hungarian			
		L1	L2	L3	L4	L1	L2	L3	L4	L1	L2	L3	L4
$\mathbf{f}_{10}$	P	52.6	52.8	52.8	52.8	34.7	34.6	34.6	34.6	46.7	49.6	49.6	49.6
	R	86.0	86.2	86.3	86.4	58.3	58.0	58.1	58.0	81.3	81.3	81.2	81.2
$\mathbf{f}_{11}$	P	32.4	32.4	32.5	32.5	32.3	32.3	32.4	32.5	30.4	30.2	30.3	30.2
	R	53.3	53.2	53.4	53.4	52.8	52.7	52.9	53.0	50.3	49.9	50.1	49.8
$\mathbf{f}_{12}$	P	52.6	53.9	53.3	53.9	**54.4**	**53.0**	**52.8**	**53.9**	53.0	52.6	53.1	53.7
	R	89.0	90.4	89.4	90.1	**90.9**	**89.2**	**88.8**	**90.0**	85.8	85.7	86.8	87.8
$\mathbf{f}_{13}$	P	48.0	49.8	46.8	44.1	46.6	44.9	42.9	41.0	46.7	44.4	41.4	40.3
	R	79.3	81.7	78.2	74.7	77.5	74.9	72.4	70.1	77.0	73.4	40.3	67.8
$\mathbf{f}_{14}$	P	33.1	32.2	33.8	31.7	31.4	30.8	32.4	30.7	32.7	32.9	30.8	32.3
	R	54.2	52.7	55.0	52.0	50.9	49.9	52.5	49.9	53.6	53.9	50.7	53.2
$\mathbf{f}_{15}$	P	38.6	38.3	38.3	38.1	32.9	32.6	32.7	32.3	34.4	34.5	34.6	34.6
	R	62.8	62.4	62.4	61.7	53.7	53.1	53.0	52.6	57.2	57.4	57.4	57.4
$\mathbf{f}_{16}$	P	22.4	21.8	19.3	19.5	22.3	21.8	19.4	19.5	22.3	21.7	19.3	19.5
	R	38.1	37.7	37.1	37.0	38.0	37.6	37.1	37.0	38.0	37.6	37.1	37.0
$\mathbf{f}_{17}$	P	23.1	21.3	22.5	22.3	27.3	24.8	25.5	26.6	30.9	28.4	29.0	29.7
	R	43.4	38.9	40.8	38.8	52.2	50.0	48.9	49.4	57.0	51.6	53.2	54.9

in two strategies, respectively. However, considering their best recall and precision scores, $\mathbf{f}_{12}$ and $\mathbf{f}_{10}$ are very close to each other, though $\mathbf{f}_{12}$ with DP is slightly better.

For the rich CS descriptors, similar to signature-based descriptors, matching algorithms have big influences on recall and precision scores (Table 6.10 and Table 6.11). Compared to other CS rich descriptors, Point Triangle ($\mathbf{f}_{19}$), Contour Context ($\mathbf{f}_{20}$) and Length Direction ($\mathbf{f}_{25}$) are stable for CS length changing and also have good matching performances (more than 90% recall). Among all these descriptors, Point Triangle ($\mathbf{f}_{19}$) with dynamic programming achieves the best performance (95.4% recall). For the Line Segment ($\mathbf{f}_{26}$) descriptor (Table 6.11) which uses the vector distance methods, its matching performance is not as good as Point Triangle ($\mathbf{f}_{19}$) with DP.

Comparing the performance of open curve matching using simple, signature-based and rich CS descriptors on the ETHZ CS dataset, the following observations can be drawn: (1) Most rich CS descriptors outperform signature-based CS descriptors. (2) Most signature-based CS descriptors perform better than the simple CS descriptors. (3) For most signature-based CS descriptors, the point matching strategy performs better than the vector distance strategy for open curve matching. (4) For open curve matching on the ETHZ dataset, Point Triangle ($\mathbf{f}_{19}$) with DP achieves the best performance (95.4% recall). (5) Considering

Table 6.9: Open curve matching results (%) with signature-based CS descriptors using vector distance on the ETHZ CS dataset.

Descriptors		Correlation				HI				χ^2-Statistics				Hellinger			
		L1	L2	L3	L4	L1	L2	L3	L4	L1	L2	L3	L4	L1	L2	L3	L4
$\mathbf{f}_{10}$	P	49.0	49.6	49.5	49.7	29.7	27.8	28.5	27.0	**53.3**	**53.3**	**52.5**	**52.4**	49.4	49.4	49.2	48.8
	R	83.6	84.3	84.2	84.4	48.9	45.6	46.8	44.9	**90.4**	**90.4**	**89.5**	**89.4**	84.3	84.1	84.0	83.6
$\mathbf{f}_{11}$	P	5.8	6.0	6.0	6.0	17.0	17.2	17.1	17.1	35.1	34.9	35.0	35.0	49.4	49.4	49.2	48.8
	R	9.9	10.2	10.2	10.3	27.9	28.2	28.0	28.1	57.6	57.3	57.5	57.5	84.3	84.2	84.0	83.5
$\mathbf{f}_{12}$	P	12.7	15.1	14.1	14.0	4.5	5.7	5.9	6.2	51.4	49.9	49.8	49.8	43.4	39.0	38.9	38.0
	R	17.2	20.5	19.3	19.3	6.6	8.3	8.6	9.1	84.4	82.8	82.7	82.9	74.8	69.3	69.1	67.8
$\mathbf{f}_{13}$	P	4.7	5.1	6.7	7.9	45.4	29.2	11.5	9.7	47.4	42.0	34.1	30.2	53.2	55.2	52.1	50.0
	R	8.2	9.2	11.5	13.0	76.7	52.5	23.3	20.3	77.4	67.9	56.4	50.7	87.8	88.8	84.3	81.2
$\mathbf{f}_{14}$	P	4.3	4.6	3.6	5.4	14.0	14.0	14.1	14.3	38.3	38.1	37.8	37.5	49.1	47.3	47.2	47.2
	R	7.7	7.8	5.6	9.5	23.0	23.1	23.2	23.4	62.6	62.4	62.0	61.6	82.9	81.2	81.0	80.6
$\mathbf{f}_{15}$	P	9.0	9.3	9.0	8.4	27.5	26.8	27.1	26.6	40.2	39.3	39.3	39.2	43.5	43.9	43.9	44.8
	R	15.3	15.7	15.3	14.3	47.6	46.7	47.0	45.9	66.6	65.4	65.4	65.0	73.2	73.7	73.7	74.9
$\mathbf{f}_{16}$	P	22.2	21.4	19.3	19.6	18.5	18.5	18.6	18.6	22.0	21.8	19.3	19.5	18.0	18.5	18.7	18.7
	R	37.8	37.0	37.1	37.1	36.7	36.6	36.9	36.9	37.2	37.3	37.0	37.0	35.4	36.1	36.9	36.9
$\mathbf{f}_{17}$	P	33.7	26.0	25.8	29.1	34.6	28.6	29.7	29.9	19.5	15.8	16.0	21.1	33.3	26.5	24.4	24.3
	R	59.7	47.0	48.0	54.7	65.6	55.6	57.7	57.9	36.0	27.2	28.4	39.6	63.0	52.1	47.8	47.0

the CS descriptors with the best performance in Table 6.2- 6.11, it can be observed that the precision and recall values of the best performances in those tables are close to each other. The main reason is that for open curve matching tasks, the influence of an individual descriptor is relatively reduced because the open curve similarity value is calculated based on the statistics of plentiful CS lengths and distances. Moreover, it can also be observed that for most CS descriptors, if one has a good performance for CS matching in the MPEG7 CS dataset, it also achieves promising results for open curve matching in the ETHZ CS dataset. Thus, signature-based and rich descriptors with proper matching algorithms can fulfill different requirements in terms of speed and accuracy for open curve matching.

6.1.1.4 Experiment 3: Computation Complexity

Table 6.12 illustrates the theoretical analysis of computational complexity for each CS descriptor. All simple CS descriptors have the same feature generation complexity $O(N)$ since they are generated by simply accumulating values computed for N CS points. As simple CS descriptors are just scalar values, their distance can be computed by scalar subtraction. Thus, its computation cost is $O(1)$.

Table 6.10: Open curve matching results (%) with rich CS descriptors and point matching methods on the ETHZ CS dataset.

Descriptors		DTW				DP				Hungarian			
		L1	L2	L3	L4	L1	L2	L3	L4	L1	L2	L3	L4
$\mathbf{f}_{18}$	P	40.8	40.3	40.0	39.3	46.0	46.5	46.5	46.3	46.8	47.2	47.2	47.2
	R	77.6	76.8	76.0	75.1	87.5	88.6	88.5	88.1	88.3	89.0	89.2	89.0
$\mathbf{f}_{19}$	P	43.0	41.1	41.0	41.0	**51.2**	**50.4**	**50.2**	**50.0**	51.0	50.1	50.0	50.0
	R	83.5	81.0	80.2	80.0	**95.4**	**94.2**	**94.0**	**94.0**	94.5	94.0	93.6	94.0
$\mathbf{f}_{20}$	P	41.5	41.5	41.5	41.5	50.0	50.0	49.5	49.5	49.0	49.0	49.0	49.0
	R	79.0	78.6	78.5	78.5	94.0	94.0	94.1	94.1	93.0	93.0	93.0	93.0
$\mathbf{f}_{21}$	P	45.3	41.2	37.3	35.3	41.0	37.0	34.0	32.4	43.5	40.1	37.3	35.3
	R	86.2	79.0	72.0	67.8	78.5	71.1	65.5	62.3	83.0	76.4	71.3	67.1
$\mathbf{f}_{22}$	P	41.1	40.2	40.0	39.6	46.1	46.3	46.2	46.1	44.6	43.6	43.1	42.9
	R	76.6	75.3	74.6	74.2	87.4	88.1	87.9	87.8	83.9	82.3	81.5	81.3
$\mathbf{f}_{23}$	P	48.8	48.7	48.7	48.8	30.1	28.4	27.8	27.7	30.8	29.0	28.5	28.3
	R	90.1	90.1	90.1	90.2	59.5	57.1	56.5	56.5	60.8	58.3	57.5	57.5
$\mathbf{f}_{24}$	P	41.5	40.5	40.0	39.7	46.7	46.9	47.0	47.0	46.4	46.7	46.8	46.9
	R	77.3	75.8	75.1	74.4	88.6	89.1	89.3	89.4	87.7	88.1	88.3	88.5
$\mathbf{f}_{25}$	P	40.6	40.6	40.6	40.6	48.7	48.7	48.7	48.7	40.4	40.4	40.5	40.4
	R	76.5	76.6	76.6	76.4	90.9	91.0	91.0	91.0	76.3	76.3	76.3	76.2

Table 6.11: Open curve matching results with rich CS descriptor and vector distance methods on the ETHZ CS dataset.

Descriptors		Correlation				HI				χ^2-Statistics				Hellinger			
		L1	L2	L3	L4	L1	L2	L3	L4	L1	L2	L3	L4	L1	L2	L3	L4
$\mathbf{f}_{26}$	P	39.3	39.0	39.2	39.1	33.2	33.2	33.4	33.2	29.2	29.2	29.3	29.2	43.7	43.5	43.7	43.5
	R	73.9	73.8	74.3	74.2	61.9	61.8	62.3	61.9	55.2	55.2	55.4	55.3	80.7	80.3	80.8	80.4

For signature-based descriptors, except Triangle Area ($\mathbf{f}_{15}$) and Chord Length ($\mathbf{f}_{16}$), most descriptors take $O(N)$ complexity for feature generation since each element is calculated only using one CS point. $\mathbf{f}_{15}$ and $\mathbf{f}_{16}$ are both calculated by considering one target CS point and other reference points selected by searching the whole CS path. Therefore, their computation complexity is $O(N^2)$. For CS retrieval, with the vector distance strategy, all four distance methods have the same computation complexity $O(N)$.

With the point matching strategy, three matching algorithms have different computation complexities. More specifically, (i) the computational complexity of DTW is $O(N^2)$ because it needs to compute distances for all possible point pairs in two CSs, each having N points. (ii) For solving the sequence alignment problem using DP, the time complexity is reduced

Table 6.12: Theoretical analysis of computation complexity for each CS descriptor in terms of feature generation and matching. NULL means this part is not considered in experiments.

Name	Feature Generation		Difference Value		
$\mathbf{f}_1$ - $\mathbf{f}_9$	$O(N)$		$O(1)$		
Name	Feature Generation	Matching (Point Matching)			Matching (Vector Distance)
		DTW [ACT09]	DP [Ric54]	Hungarian [Kuh55]	
$\mathbf{f}_{10}$ - $\mathbf{f}_{14}$, $\mathbf{f}_{17}$	$O(N)$	$O(N^2)$	$O(N^2)$	$O(N^3)$	$O(N)$
$\mathbf{f}_{15}$, $\mathbf{f}_{16}$	$O(N^2)$	$O(N^2)$	$O(N^2)$	$O(N^3)$	$O(N)$
$\mathbf{f}_{18}$ - $\mathbf{f}_{20}$ $\mathbf{f}_{22}$ - $\mathbf{f}_{25}$	$O(N^2)$	$O(N^2)$	$O(N^2)$	$O(N^3)$	NULL
$\mathbf{f}_{21}$	$O(N)$	$O(N^2)$	$O(N^2)$	$O(N^3)$	NULL
$\mathbf{f}_{26}$	$O(N)$	NULL	NULL	NULL	$O(1)$

from $O(N^3)$ with the traditional brute-force approach [Ric54]) to $O(N^2)$ with the method introduced in [Sel80]. This is because the method in [Sel80] makes a complete list of all pairs of intervals using given CSs so that each pair displays a maximum local degree of similarity. Like this, the matching complexity is reduced by trading space for time. (iii) The Hungarian algorithm solves CS matching tasks in $O(N^3)$ time as introduced in [LS10].

For rich CS descriptors, it can be observed that most descriptors require (or more than) $O(N^2)$ complexity for feature generation. In contrast, Beam Angle ($\mathbf{f}_{21}$) is calculated by the angles between CS points and its neighbouring points which can be captured directly. Line Segment ($\mathbf{f}_{26}$) is calculated by the statistics of a fixed range of straight-line scales n (n is set from 5% to 85% in all experiments). Since n is independent of CS point number N and $n \ll N$, the computational complexity of this method is determined by the number of CS points. Therefore, the complexity of $\mathbf{f}_{21}$ and $\mathbf{f}_{26}$ is $O(N)$. For CS matching, the point matching strategy is applied to descriptors ranging from $\mathbf{f}_{18}$ to $\mathbf{f}_{25}$, in which the matching algorithms have the same complexity as signature-based descriptors. For the vector distance strategy, Line Segment ($\mathbf{f}_{26}$) has the complexity $O(n)$ where n is the feature dimension. Since its feature dimension is fixed and the distance between two vectors can be calculated in a constant time, its computational complexity is $O(1)$.

6.1.1.5 Discussion

Table 6.13 illustrates the comparison between the selected descriptors which have outstanding matching performances on ETHZ and MPEG7 CS datasets. It can be observed that the selected signature-based and rich descriptors have robust performances in which Beam Angle ($\mathbf{f}_{21}$) with Hungarian [Kuh55] achieves the best bulls-eye score (77.6%). Tangent ($\mathbf{f}_{12}$) and Point Triangle ($\mathbf{f}_{19}$) with DP [Ric54] obtain the best precision (53.5%) and recall

Table 6.13: Matching performance and time (hour) comparison between selected CS descriptors which have outstanding matching performances on ETHZ and MPEG7 CS datasets.

Name	Notation	Type	Method	FG	Matching	ETHZ CS*	MPEG7 CS*
Comcoor	$\mathbf{f}_{10}$	Signature	χ^2-Statistics	0.0001	0.0542	P: 52.9, R: 89.9	56.2
Comcoor	$\mathbf{f}_{10}$	Signature	Hungarian	0.0001	213.74	P: 48.9, R: 81.3	64.0
Comcoor	$\mathbf{f}_{10}$	Signature	Hellinger	0.0001	0.0531	P: 49.2, R: 84.0	63.6
Cendistance	$\mathbf{f}_{11}$	Signature	Hellinger	0.0001	0.0462	P: 49.2, R: 84.0	63.6
Tangent	$\mathbf{f}_{12}$	Signature	DP	0.0001	3.86	P: 53.5, R: 89.7	57.8
Point Triangle	$\mathbf{f}_{19}$	Rich	DP	6.82	121.8	P: 50.5, R: 94.4	69.7
Contour Context	$\mathbf{f}_{20}$	Rich	DP	0.04	6.7	P: 49.6, R: 94.1	63.7
Beam Angle	$\mathbf{f}_{21}$	Rich	Hungarian	0.01	147.1	P: 39.1, R: 74.5	77.6
Partial Contour	$\mathbf{f}_{22}$	Rich	DP	0.18	5.3	P: 46.2, R: 87.8	74.1
Opt Partial Contour	$\mathbf{f}_{23}$	Rich	DTW	0.17	22.1	P: 48.8, R: 90.1	60.0
Chord Distribution	$\mathbf{f}_{24}$	Rich	DP	0.18	5.4	P: 46.9, R: 89.1	74.6
Length Direction	$\mathbf{f}_{25}$	Rich	DP	0.02	212.1	P: 48.7, R: 91.0	64.3

* As there are four values for four different lengths, the mean value is calculated for comparison.
P: Precision, R: Recall, FG: Feature Generation

(94.4%). However, considering the runtime and retrieval performance, Partial Contour ($\mathbf{f}_{22}$) and Chord Distribution ($\mathbf{f}_{24}$) are close to the best while taking less time for feature generation and matching.

Table 6.14: Retrieval Results on two datasets using the fused descriptors. D-Value denotes the difference between the two values.

Descriptors	Matching Algorithm	ETHZ CS Dataset		MPEG7 CS Dataset
		Precision	Recall	
Eccentricity ($\mathbf{f}_3$)	D-Value	55.1	90.1	22.7
Point Triangle ($\mathbf{f}_{19}$)	DP	50.5	94.4	69.7
Fused	DP + D-Value	56.1	97.6	74.6

To obtain a state-of-the-art performance for open curve matching on a real-world dataset, multiple CS descriptors should be chosen and fused [SGS10]. As discussed in [Yan+14b], even a combination of simple shape descriptors improves the overall performance of individual descriptors. According to the matching performance on ETHZ and MPEG7 CS datasets, Eccentricity ($\mathbf{f}_3$), which is the best simple CS descriptor on both datasets, is fused with other signature-based and rich CS descriptors. To compute proper fusing weights, the dataset is firstly divided into two equal parts, one used for weight estimation and the other for testing. For fusion weight estimation, a supervised optimisation scheme [Yan+15b] is employed in which two heuristic approaches are combined. The matching performance is experimentally assessed by fusing the Point Triangle ($\mathbf{f}_{19}$) and Eccentricity ($\mathbf{f}_3$) descriptors. The experiments show that, compared to Point Triangle ($\mathbf{f}_{19}$), the fused descriptor improves the matching accuracy by 4.4% on the ETHZ CS dataset and by 4.9% on the MPEG7 CS dataset (Table 6.14). Therefore, a proper combination of CS descriptors can improve the matching accuracy over the individual descriptors.

6.1.2 Shape Contour Detection

In order to evaluate the shape contour detection method introduced in Section 2, one experiment is applied in this part using the ETHZ shape classes dataset [FJS09]. As introduced in Section 6.1.1, this dataset features five diverse classes (apples, bottles, giraffes, mugs and swans) and contains a total of 255 images collected from the Internet. It is highly challenging, as the objects appear in a wide range of scales; there is considerable intraclass shape variation, and many images are severely cluttered. Based on this dataset, two groups of shapes are generated. In the first group, shapes are generated by the ground truth. In the second group, shapes are generated by the shape contour detection method in Section 2. Some sample shapes and their ground truths are shown in Figure 6.3.

Figure 6.3: Sample shapes from the ETHZ [FJS09] database. The shapes in the first and third row are the ground truths which are manually generated. The shapes in the second and fourth row show the segmented shapes using the shape contour detection method in Section 2.

It can be clearly observed that the generated shapes visually cover the most region of their ground truths. However, there are still some regions missed since only stable open curves are returned during the open curve reduction process. In such a case, some regions are excluded due to the loss of small open curves. In order to solve this problem, on the one hand the correlated parameters can be modified to include more open curves. On the other hand, multiple segmentation methods can be fused to improve the stability of shape contour detection like [BST15]. To quantitatively assess the contour detection performance, some standard evaluation frameworks like Berkeley segmentation benchmark [DRB10a] can be applied on the generated shapes. However, as the generated shape will be used to compare and evaluate the performance of shape matching methods in Section 6.3.1.5, there is no need to specially apply additional experiments in this part since the evaluated results in Section 6.3.1.5 can be regarded as evidence of the usability for object shape generation.

6.2 2D Object Retrieval using Coarse-grained Features

In this part, shape retrieval and classification experiments are reported using the proposed coarse-grained shape descriptors and their correlated matching and classification methods. Experiments in this section are performed on a laptop with Intel Core i7 2.2GHz CPU, 8.00GB memory and 64-bit Windows 8.1 OS. All methods in the experiments are implemented in Matlab R2015a.

6.2.1 Object Retrieval using the Contour-based Method

To evaluate the proposed methodology, two datasets are used in the shape-based object retrieval scenario:

(1) Kimia216: The Kimia216 [SKK04] dataset contains 216 shapes from 18 classes. Figure 6.4 shows two example shapes in each of these 18 classes.

Figure 6.4: Sample shapes from the Kimia216 [SKK04] database.

(2) MPEG400: The MPEG400 dataset is a subset of the MPEG7 [LLE00] dataset, consisting of 400 objects categorised in 20 classes. Compared to MPEG7, shapes in MPEG400 have much larger intra-class variations and inter-class similarities (Figure 6.5).

Figure 6.5: Sample shapes from the MPEG400 [Yan+14b] database.

Table 6.15: Experimental comparison of the proposed methodology to the most powerful related algorithm using Kimia216 and MPEG400 datasets. Results are summarised as the number of shapes from the same class among the first top 1-11 shapes.

Kimia216	1st	2nd	3rd	4th	5th	6th	7th	8th	9th	10th	11th
PSSGM [BL08]	216	216	215	216	213	210	210	207	205	191	177
Revised PSSGM [Hed+13]	205	208	202	199	200	192	184	167	161	130	96
Proposed Method	**216**	**215**	**206**	**204**	**200**	**186**	**172**	**163**	**130**	**124**	**107**

MPEG400	1st	2nd	3rd	4th	5th	6th	7th	8th	9th	10th	11th
PSSGM [BL08]	380	371	361	351	344	339	332	320	330	309	305
Proposed Method	**375**	**348**	**333**	**325**	**317**	**311**	**300**	**295**	**276**	**275**	**259**

Based on Kimia216 and MPEG400 datasets, the proposed method is compared with the PSSGM [BL08] method which has the state-of-the-art performance for shape retrieval. During the experiment, each shape in a dataset is used as a query and the retrieved results are checked whether they are correct, i.e. belong to the same class as the query. In order to enable a quantitative comparison, the experimental convention proposed in [BL08] is kept so that the 11 best matches are considered for each query. Results achieved from Kimia216 and MPEG400 datasets can be found in Table 6.15, whereas in [Hed+13] the PSSGM algorithm has been re-evaluated without any preliminary assumptions regarding the object skeletonisation.

Based on the results in Table 6.15, it can be clearly observed that the proposed open curve-based shape descriptor together with the similarity measure is very robust in both datasets, since it leads to very good shape retrieval results without using any additional discriminative properties of a shape. Moreover, though the proposed shape descriptor is built on the coarse-grained features, it still has a promising performance since its retrieval results are already close to the PSSGM method which is composed by fine-grained features.

In Section 6.4.1 and 6.4.2, two additional experiments are conducted with fused descriptors to evaluate the improvement of fine-grained descriptors that are fused with the proposed coarse-grained descriptor.

6.2.1.1 Computational Complexity

Here the computational complexity of the feature generation and matching approaches in Section 3.2 and 5.2.1 is analysed. (1) For feature generation, the shape descriptor in Section 3.2 generally includes two parts: DCE-based shape boundary partition and feature generation for each open curve. For shape boundary partition, the time complexity is $O(v' \log v')$, where v' is the number of the vertices in the original polygon P. This is because the DCE method also has the time complexity $O(v' \log v')$ [BLL07]. After partition, for an open curve C', the time complexity for feature generation is $O(l(C'))$, where $l(C')$ is the length of C'. Assume there are N open curves generated a given shape boundary ∂D, the total complexity is $O(v' \log v') + N \cdot O(l(C'))$. Recalling that $N \ll v'$ and $O(l(C'))$ can also be finished in a constant time, the total time complexity for descriptor generation is bounded by $O(N' \log N')$. (2) For shape matching, the time complexity for generating the similarity matrix $\mathbf{S}(D^\star, D^\diamond)$ is $O(N_1 N_2)$, where N_1 and N_2 are the number of open curves in shape $D^\star$ and $D^\diamond$, respectively, $N_1 > N_2$. Based on the similarity matrix $\mathbf{S}(D^\star, D^\diamond)$, a Hungarian algorithm is applied with the time complexity $O(N_1^3)$. Thus, the total time complexity for shape matching is $O(N_1 N_2) + O(N_1^3)$. Since both $N_1, N_2 \ll v'$, a shape matching task can be finished in $O(1)$ time complexity.

6.2.2 Object Retrieval and Classification using the Region-based Method

In this part, the proposed region-based shape descriptor with supervised optimisation methods is evaluated via two scenarios: shape retrieval and classification.

6.2.2.1 Shape-based Object Retrieval

First, the object retrieval experiment on the MPEG7 [LLE00] dataset is reported. As introduced in Section 6.1.1, the MPEG7 dataset has $70 \times 20 = 1{,}400$ shapes (Figure 6.6). In order to optimise the parameters involved in the proposed matching algorithm, 10 objects from each category are randomly selected and there are in total $10 \times 70 = 700$ objects used for supervised optimisation. After optimising all parameters, the remaining 700 objects are employed for testing. Table 6.16 shows the retrieval results with different optimisation options.

In order to quantitatively evaluate the results on the MPEG7 dataset, the bulls-eye score is employed for result analysis. Every shape in the dataset is compared to all other shapes,

Figure 6.6: Sample shapes from the MPEG7 [LLE00] database.

and the number of shapes from the same class among the 20 most similar shapes is reported. The bulls-eye scores from other references are on all 1400 shapes, while here only 700 shapes are used for testing. Therefore, in order to ensure the correctness of the analysis, the training/testing processes are applied multiple times by iteratively splitting the dataset. Finally, an average value is reported. The proposed supervised optimisation with the matching algorithm achieves a 94.0% bulls-eye score on the MPEG7 dataset.

Although Michael [DB13] has already achieved a 100% bulls-eye score on the MPEG7 dataset, the purpose of their method is different from the proposed one. In this part, a simple

Table 6.16: Retrieval results on the MPEG7 dataset. Results are summarised as the number of shapes from the same class among the first top 1-10 shapes. No Opt shows results from the proposed matching algorithm without optimisation. Opt1 shows results using the proposed matching algorithm with Graph Transduction optimisation. The last row shows results using the proposed matching algorithm with the supervised optimisation method.

	1st	2nd	3rd	4th	5th	6th	7th	8th	9th	10th
No Opt	700	647	600	567	521	488	447	426	405	342
Opt1[Bai+10]	640	584	552	501	463	424	398	381	303	145
Proposed Method	**700**	**657**	**615**	**591**	**553**	**518**	**475**	**467**	**420**	**363**

and effective coarse-grained shape descriptor is proposed that can be fused with other descriptors. In order to ensure its flexibility to equip it with a higher adaptiveness, a supervised optimisation strategy is introduced. The proposed method does not involve any manifold structure defined by pairwise affinity matrices. However, Graph Transduction [Bai+10] and diffusion [DB13] can be used in the last step to improve subsequent applications like object retrieval.

In order to fully assess the quality of the proposed method, the second experiment is applied based on the MPEG400 dataset since shapes in this dataset have much larger intra-class variations. The object retrieval results are compared to the contour-based method in Section 6.2.1. Since this database is the subset of MPEG7, the same parameter values from the first experiment are employed to implement the object retrieval experiment on MPEG400. To evaluate the behaviour of the proposed descriptor and the matching algorithm, the experiments are applied in both with and without optimisation settings. As shown in Table 6.17, experiment results show that the proposed method performs better than the contour-based method. Moreover, it can be clearly observed that the proposed matching algorithm with supervised optimisation made a significant progress on the MPEG400 dataset.

Table 6.17: Object retrieval on the MPEG400 dataset. The second row represents the retrieval results using the proposed matching algorithm with default parameters; the last row shows the retrieval results using the proposed matching algorithm with optimised parameters.

	1st	2nd	3rd	4th	5th	6th	7th	8th	9th	10th
Contour-based[Yan+14b]	375	348	333	325	317	311	300	295	276	275
Proposed Method (No-Opt)	**381**	**355**	**341**	**320**	**322**	**316**	**304**	**295**	**269**	**260**
Proposed Method (Opt)	**381**	**370**	**365**	**354**	**337**	**342**	**328**	**315**	**300**	**301**

6.2.2.2 Shape-based Object Classification

In this part, the experiments of shape-based object classification are implemented based on the MPEG7 dataset. As already discussed in Section 5.2.2.2, here the RBF is used as the kernel function for SVM. Like in the previous experiments, shapes in the MPEG7 dataset are randomly split into one half for training and another half for testing. The training set should include all object classes. During the training phase, there are two main tasks: (1) building classifiers and (2) seeking best parameters with the proposed optimisation strategy. All of these tasks are done by 700 training shapes. In this experiment, there are two parameters that need to be optimised, C for the cost of SVM and γ in the RBF kernel function. The default values are $C = 1$ and $\gamma = 1/n$, where n is the number of features. Since there are 10 feature values in the proposed feature vector, $n = 10$ and $\gamma = 0.1$.

Table 6.18: Object classification for the whole MPEG7 dataset. The first row represents the results of scaled datasets with default parameters. The proposed method achieves significant progress in this experiment.

	1st	2nd	3rd	4th	5th	6th	7th	8th	9th	10th	Average
Default(%)	47.6	45.6	50.7	47.6	46.0	44.4	46.4	46.0	45.9	44.9	46.5
Proposed Method(%)	**85.7**	**86.0**	**85.9**	**85.7**	**84.1**	**86.6**	**85.1**	**87.7**	**87.9**	**88.0**	**86.3**

In order to increase statistical relevance, the selection process is repeated 10 times which led to 10 different training sets and their correlated testing sets. As shown in Table 6.18, each column refers to an experiment with different training datasets. Experiments are performed for all these datasets and mean classification rates are reported. Results shows that the proposed method achieves significant progress in this experiment. Essentially, SVM and RBF are highly related to parameter selection and the proposed supervised optimisation method can improve their performances sufficiently.

Table 6.19: Experimental comparison of the proposed methodology to the related algorithm for object classification on the MPEG7 dataset.

Method	Score
Skeleton Path[BLT09]	86.7%
Class Segment Sets[SS05]	75.4%
ID[LJ07]	76.5%
Proposed Method	**86.3%**

Table 6.19 reports the comparison between the proposed method and some existing classification methods on the MPEG7 dataset. It can be observed that the proposed method yields a promising score compared to other methods. Though the Skeleton Path [BLT09] approach achieves similar results as the proposed one, the computation complexity of the proposed shape descriptor is much lower than that of the skeleton path which needs to involve skeleton pruning for feature generation. It should be noted that a descriptor in [BLT09] achieved 96.6% classification rate on the MPEG7 dataset. However, this method cannot be directly used for comparison with the other method in Table 6.19 since the descriptor in [BLT09] is fused with two other fine-grained shape descriptors. In contrast, the proposed descriptor is only composed by coarse-grained shape features which can be easily fused with other fine-

grained descriptors. In Section 6.4.3, one example is illustrated to prove the improvements of integrated descriptors using the proposed one.

6.2.2.3 Computational Complexity

In this part, the computational complexity for the shape descriptor in Section 3.3, shape matching and classification methods in 5.2.2 are illustrated. (1) For shape generation, the first feature is generated by the time complexity $O(N')$, where N' is the length of the shape boundary ∂D. For the last feature, since a shape skeleton is generated by the fast thinning algorithm in [ZS84], the time complexity is $O((N'')^2)$, where N'' is the number of pixels in a shape D. The remaining features are generated by ratios between elements in different bounding boxes with the time complexity $O(1)$. Thus, the total time complexity for feature generation is $O(N') + O((N'')^2)$. As $N' \ll N''$, the total time complexity is bounded by $O((N'')^2)$. (2) For shape matching, the time complexity is $O(1)$ since the dimensionality of a feature vector in Equation (5.7) is fixed. For shape classification, the time complexity for training SVM takes $O(\max(N_{samples}, N_{features}) \min(N_{samples}, N_{features})^2)$ [Cha07], where $N_{samples}$ is the number of samples and $N_{features}$ is the number of dimensions. As illustrated in Section 3.3, $N_{features} = 10$ and the number of training sets is normally larger than $N_{features}$. Thus, the time complexity is $O(N_{samples} N_{features}^2)$.

6.3 2D Object Retrieval using Fine-grained Features

In this part, the experiments using two proposed fine-grained shape descriptors and their matching algorithms are introduced. Experiments in Section 6.3.1 are performed on a laptop with Intel Core i7 2.2GHz CPU, 8.00GB memory and 64-bit Windows 8.1 OS. In addition to the laptop, some experiments in Section 6.3.2 are also performed on a cluster. Specifically, hierarchical skeletons are generated on the laptop introduced above. Shape retrieval experiments are accomplished on *Horus* which is introduced in Section 6.1.1.1. All methods in the experiments are implemented in Matlab R2015a.

6.3.1 Object Retrieval using the Contour-based Method

In this section the proposed interest point detector and point context on different datasets are firstly evaluated. After that, the performance of the proposed matching method with some traditional methods is evaluated to illustrate the addressed advantages. Lastly, the proposed method is compared to the related ones on three datasets.

6.3.1.1 Evaluation of Interest Point Detection

To evaluate the effectiveness of the proposed interest point detector, a standard dataset Kimia216 [SKK04] is employed. The evaluation is based on the retrieval framework, where each shape is used as a query, and the 10 most similar shapes are retrieved from the whole dataset. The performance of a method is evaluated by checking retrieval results using all 216 shapes as queries, and counting how many retrieved shapes belong to the same class of their queries. The similarity value between each shape and its query is calculated using the interesting point matching method in Section 5.3.1.2. Since the performance of interest point detection is tightly coupled with the end goal of matching, the similarity values are calculated and compared by using the same point descriptors and the matching algorithm. Built on this, the influence of the point descriptor and matching algorithm can be minimised by keeping the only variable: point detection methods. Thus, the experiments are more targeted to the performance of interesting point detection.

The proposed interesting point detection method is compared to the most related DCE [LL99] and corner detection [Liu+07]. The DCE method detects interesting points by considering the polygon convex during the iterative polygon simplification. In the Corner detection method, a contour point is described by both its visual curvatures and corresponding scales. In a certain scale, they consider the points of which their digital visual curvature is above a threshold DK_0 as interesting points. Based on the generated interesting points, retrieval performances of the proposed method and the other two methods are compared using the same descriptors, SC [BMP02] and the proposed point context descriptor. As discussed in Section 4.1, DCE requires a stop parameter k to control polygon simplification. For a fair comparison, several parameters are tried from $k = 3$ to $k = 15$ and finally $k = 10$ is chosen since it achieved the best retrieval result for comparison. In the corner detection method, their mentioned threshold $DK_0 = 17\pi/64(48°)$ is employed.

Table 6.20 presents the performance comparison between the interesting point detection method and two other methods. The results in this table are collected by checking retrieval results using all the 216 shapes as queries. For example, the fourth position in the row of IP1 shows that from 216 retrieval results in the forth position, 184 shapes belong to the same class as their query shapes. It can be observed that retrieval results based on the proposed interesting point detector perform better on both descriptors. The main reason is that the property of interesting points generated by the DCE method is highly related to the stop parameter k. Obviously, it is impractical to set an appropriate k manually on each object. Similar to DCE, the corner detection method also detects the interesting points based on the visual curvature which is higher than the threshold DK_0. Since this threshold is not general for all the shapes in this dataset, some interesting points could be mis-detected. On the contrary, the proposed method can generate stable points with no sensitive parameter.

Table 6.20: Experimental comparison on the Kimia216 dataset. SC: Shape Context [BMP02] descriptor, PC: The proposed point context descriptor, IP1: Interesting points detected by the DCE method [LL99], IP2: Interesting points detected by the visual curvature method [Liu+07], IP3: Interesting points detected by the proposed method.

SC	1st	2nd	3rd	4th	5th	6th	7th	8th	9th	10th
IP1	216	210	195	184	181	172	161	146	148	128
IP2	216	205	195	190	187	179	180	170	171	161
IP3	**216**	**212**	**206**	**197**	**191**	**190**	**186**	**186**	**183**	**171**

PC	1st	2nd	3rd	4th	5th	6th	7th	8th	9th	10th
IP1	216	211	205	196	192	191	186	178	177	175
IP2	216	210	205	203	194	188	179	170	160	155
IP3	**216**	**212**	**211**	**211**	**205**	**200**	**201**	**195**	**193**	**195**

6.3.1.2 Evaluation of Point Context

In order to evaluate the performance of the point context descriptor, the retrieval performance is compared to other most related descriptors, SC [BMP02] and Extended Shape Context (PCCS) [Fei+14]. For an interesting point, the SC [BMP02] method extracts descriptors as the diagram of the bins which are uniformed in log-polar space. The PCCS method considers interesting points as the contour partition points; then, the shape is represented as the fused SC descriptor on interesting points and contour segment descriptors. In this experiment, the MPEG400 dataset which consists of 400 objects categorised in 20 classes is used. Except for the descriptors, all performances are obtained using the same interest points based on our proposed method and are matched by the Hungarian algorithm.

Table 6.21 illustrates the experimental comparison between three descriptors and the point context descriptor achieves the best performance. The presentation format of Table 6.21 is the same as the one of Table 6.20, except for the fact that 400 query shapes are used in Table 6.21. It is clear that PCCS performs better than SC, since PCCS considers not only shape contexts on interesting points, but also the geometrical features on contour segments. However, compared to PCCS, point context performs better since it is generated by taking both distance and orientation features for measuring the distribution of relative positions from an interesting point to the sample points.

Table 6.21: Experimental comparison of the proposed point context descriptor to the SC [BMP02] and PCCS [Fei+14] using the interesting points generated by the proposed method on the MPEG400 dataset. The matching algorithm on all the three methods is the Hungarian algorithm.

	1st	2nd	3rd	4th	5th	6th	7th	8th	9th	10th
SC [BMP02]	370	343	310	302	277	272	265	264	239	240
PCCS [Fei+14]	377	351	336	331	317	302	287	282	273	262
Proposed Method	**391**	**377**	**372**	**364**	**356**	**343**	**338**	**319**	**304**	**276**

6.3.1.3 Evaluation of High-Order Matching

In this section, performances of the proposed high-order graph matching method are quantitatively evaluated. First of all, the performance of high-order graph matching is visually compared to the traditional Hungarian method [Kuh55] and the state-of-the-art method in [Fei+14] using the same objects. In Figure 6.7, two hands are matched with deformations on some fingers using only the first-order potentials (i.e. matching among single points by the Hungarian algorithm) and its combination with the third-order potentials. As shown in Figure 6.7 (a), there are some mismatched interesting points because of the similar points in both shapes. Moreover, the geometrical relations among the interesting points are not considered. Figure 6.7 (b) shows that the proposed high-order matching method yields appropriate matching. Since there are more interesting points in the left hand than the right one, with the constraint in Section 5.3.1.1, some points in the left hand will be left out.

In Figure 6.7 (b), it can be clearly observed that an interesting point in the right hand (the bottom right corner) is not properly detected. This is because interesting points are detected by considering both curvatures and the overall shape tend. If a corner point and some contour points have similar distances to a reference point, the corner point could be removed during the sequence smoothing steps. Because of this, it can be observed that the interesting point in the left hand (the bottom right corner) could not be found as properly corresponding. As a result, it is assigned to another point to meet the singleton and high-order constraints. It also influences the correspondences of its adjacent points. Thus, the proposed method has erroneous correspondences around the wrist area. This problem could be solved by enriching the type of interesting points and setting up a threshold to remove the ambiguous correspondences.

In Figure 6.8, the matching results in [Fei+14] are compared to the proposed high-order matching approach. Depending on human perception, there are several mismatched points in Figure 6.8 (left). The main reason is their symmetric silhouette which renders several points difficult to match based on single-point matching. In contrast, as shown in Figure 6.8

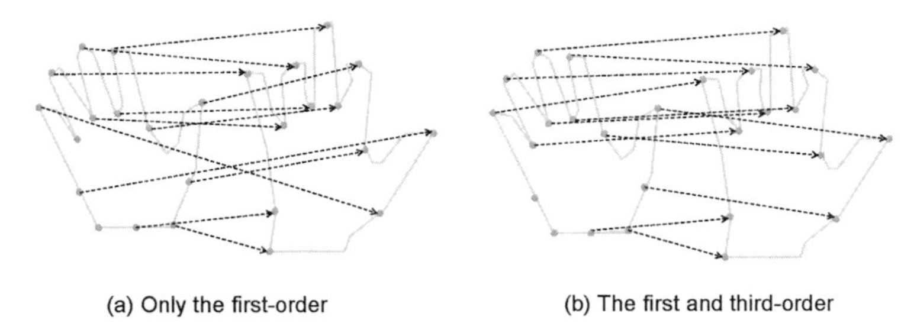

(a) Only the first-order (b) The first and third-order

Figure 6.7: Object matching with different potentials. It can be observed that the high-order potential performs better.

(right), with the point context feature and the proposed matching approach, all points in the left tool are correctly matched to the right one.

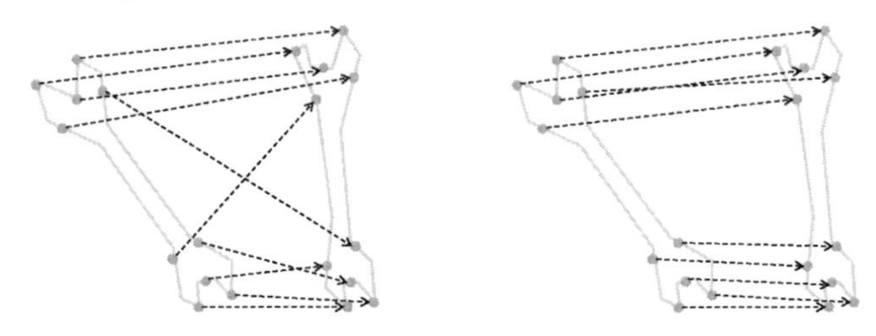

Figure 6.8: Comparing the matching result of the proposed method (right) to the method in [Fei+14] (left).

Next, the improvement of matching accuracy using the high-order graph matching method is quantitatively evaluated. For this, the Kimia99 [SKK04] database is used. Kimia99 contains images of 9 categories of objects, with 11 images per species for a total of 99 images (Figure 6.9). Table 6.22 depicts the comparison between incorrect correspondences found using only the first-order potentials (i.e. using the Hungarian algorithm) and those found using the higher-order methods. Using interesting points detected by the proposed method, the total number of correspondences between shapes in the same class is firstly counted. The values in the second column in Table 6.22 illustrate the total number of correspondences in each class. After that, the Hungarian and different high-order matching methods are applied independently and the incorrect correspondences between shapes in the same class are collected. The values in the third and the last column present the total numbers of incorrect correspondences using the Hungarian and the proposed method, respectively. The values in the fourth and the fifth column are generated using the Hyper-

Graph Matching (HGM) [ZS08] and the Reweighted Random Walks-based Hyper-Graph Matching (RRWHM) [LCL11].

Figure 6.9: Sample shapes from the Kimia99 [SKK04] dataset.

It can be observed that the RRWHM method achieves a performance close to the proposed approach while the HGM has the lowest performance among the three high-order matching methods. The main reason is that the HGM method is unable to effectively incorporate with the matching constraints during its approximation stage. Different from HGM, the RRWHM method works with the mapping constraints in the approximation stage which can effectively reflect the one-to-one matching constraints during the random walks for higher-graph matching. Overall, Table 6.22 indicates that high-order matching methods can significantly improve the shape matching performance and reduce the number of incorrect matches.

Table 6.22: Experimental comparison of incorrect correspondences between the Hungarian and high-order matching methods in each class.

Class	Total	Hungarian	HGM	RRWHM	**Proposed Method**
animal	1070	270	261	166	**176**
bunny	852	115	84	4	**14**
dude	1034	157	178	83	**94**
fish	414	49	62	27	**28**
hand	1319	497	286	180	**180**
hat	712	193	140	78	**82**
key	947	1	0	0	**0**
plane	1120	450	292	213	**203**
tool	718	130	85	18	**26**

6.3.1.4 Performance Comparison to State-of-the-art Methods

Kimia99 Database: Table 6.23 shows the performance comparison between the proposed method and the most famous methods like ID [LJ07], SC [BMP02] and PSSGM [BL08]. These methods are commonly used for shape matching with different descriptors. It can be clearly observed that the proposed method performs better than the ID and SC methods while being close to the best results from the PSSGM method in this database. The main reason is that skeletons employed in the PSSGM method are perfectly pruned with human interaction. In contrast, the whole process of the proposed method can be performed without any human interaction.

Table 6.23: Experimental comparison of our method to ID [LJ07], SC [BMP02] and PSSGM [BL08] on the Kimia99 dataset. The last column illustrates the overall feature generation and matching time in hours.

	1st	2nd	3rd	4th	5th	6th	7th	8th	9th	10th	Time
ID [LJ07]	99	97	92	89	85	85	76	75	63	53	0.15
SC [BMP02]	99	97	91	88	84	83	76	76	68	62	0.69
PSSGM [BL08]	99	99	99	99	96	97	95	93	89	73	0.38
Proposed Method	**99**	**99**	**96**	**92**	**88**	**84**	**80**	**78**	**73**	**60**	**0.36**

Tetrapod120 Database: The Tetrapod120 database is specially organised for this dissertation, includes 120 visually similar tetrapod animals with 6 classes, such as camel, cattle, deer, dog, elephant and horse. Figure 6.10 shows example shapes in Tetrapod120 and it can be seen that some species are really difficult to distinguish by their shapes.

Figure 6.10: Sample shapes from the Tetrapod120 database.

With this dataset, the aim is to evaluate the ability of matching methods for fine-grained shapes. In other words, the ability for handling local shape deformations is assessed using this dataset. As illustrated in Table 6.24, the proposed method achieves the best results. This indicates that the geometric relationship between interesting points is an important feature for distinguishing fine-grained objects.

MPEG7 Database: Similar to previous sections, the bulls-eye score [LLE00] is employed in this dataset for evaluation. For comparison, Table 6.25 lists several reported results

Table 6.24: Experimental comparison of the proposed methodology to ID [LJ07], SC [BMP02] and PSSGM [BL08] using the Tetrapod120 dataset. The last column illustrates the overall feature generation and matching time in hours.

	1st	2nd	3rd	4th	5th	6th	7th	8th	9th	10th	Time
ID [LJ07]	120	118	106	101	90	83	77	69	70	56	0.22
SC [BMP02]	100	80	70	53	53	51	40	28	27	27	0.96
PSSGM [BL08]	120	109	101	98	81	78	68	66	65	59	1.41
Proposed Method	**120**	**115**	**111**	**105**	**105**	**103**	**98**	**93**	**94**	**87**	**0.53**

and the results by the proposed method on the MPEG7 dataset. As reported in [Bai+12], existing methods are mainly clustered into two groups: pairwise matching and context-based matching. In the first group, results are decided by the similarity measures for shape pairs. In the second group, results are generated by considering the underlying structure of the shape manifold [Bai+12] in which the obtained similarity scores are post-processed by analysing the shape similarities between all given shapes to increase the discriminability between different shape groups.

In the first group, the proposed matching method achieves a 80.28% bulls-eye score which is better than a traditional contour-based descriptor [BMP02]. However, this approach performs not as well as Shape Tree [FS07] and HF [Wan+12a], etc. The main reason is that only 240 sample points from a shape contour (containing more than 1500 points) are used to generate point context features for interesting points and the performance can be improved by using more sample points. In the group of context-based methods, the MG method [KDB10] is used based on the proposed similarity scores between all the shapes. The proposed method, which achieves 96.43% bulls-eye score, outperforms most state-of-the-art methods. It is important to mention that Donoser *et al.* [DB13] proposed a generic framework for diffusion processes in the scope of retrieval applications which achieved 100% accuracy on the MPEG7 dataset. However, as illustrated in Table 6.25, the performance of the proposed method comes close to 100% only using the simple Mutual *k*NN Graph method.

6.3.1.5 Experiments on Imperfect Segmented Shapes

In order to evaluate the proposed method in the case of unsegmented natural image settings, two experiments are applied using the ETHZ shape classes dataset [FJS09] introduced in Section 6.1.1. Based on this dataset, two groups of shapes are generated. In the first group, shapes are generated by the ground truth shapes. Some noise is also introduced into the shapes to mimic imperfect segmentations. In the second group, shapes are generated by

Table 6.25: Bulls-eye score on the MPEG7 Dataset.

Pairwise Matching	Score	Context-based	Score
SC [BMP02]	76.51%	INSC + CDM [Jeg+10]	88.30%
Skeletal Context [XHS08]	79.92%	IDSC + LP [Bai+10]	91.00%
Optimized CSS [MB03]	81.12%	SC + LP [Bai+10]	92.91%
Multiscale Rep. [AO04]	84.93%	IDSC + LCDP [YKTL09]	93.32%
Shape L'AneRouge [PRH08]	85.25%	SC + GM + Meta [EKG10]	92.51%
Fixed Cor. [Sup06]	85.40%	IDSC + MG [KDB10]	93.40%
ID [LJ07]	85.40%	IDSC + PSSGM + LDCP [Tem+10]	95.60%
Symbolic Rep. [DT08]	85.92%	ASC + LDCP [LYL10]	95.96%
Hier.Procrustes [MV06]	86.35%	HF + LCDP [Wan+12a]	96.45%
Triangle Area [AKF08]	87.23%	SC + DDGM + Co-T [Bai+12]	97.45%
Shape Tree [FS07]	87.70%	AIR [GTC10]	93.67%
Height Functions [Wan+12a]	89.66%	ASC + TN + TPG [YPL13]	96.47%
Proposed Method	**80.28%**	**Proposed Method + MG [KDB10]**	**96.43%**

the shape generation method introduced in Chapter 2. Some sample images and their segmented shapes are shown in Figure 6.11.

In the first experiment, the proposed point detection method is applied on two shape groups and reports the False Positive (FP) and False Negative (FN) rates of detecting interesting points in Table 6.26 (the upper table). It can be observed that the FP rates in two groups are both promising (1.66% and 0.81%) while the FN rates are much higher (11.26% and 9.64%). The main reason is that the proposed point detection method only takes the boundary points by mainly considering the overall shape tend. In this case, some corner points could be ignored which leads to the high FN rates. This problem could be solved by considering both overall shape tend and boundary corner to reduce the FN rate of the proposed method.

In the second experiment, two shape groups are used in the shape retrieval scenario. The mean accuracies within each class and the whole dataset are reported in Table 6.26 (the lower table). Though the mean accuracy in the shape group with manually added noise is higher than the shapes in another group, the overall performances of the proposed method are not promising in both shape groups. There are two reasons for this: The first reason is that the integrated high-order graph matching could mislead the partial matching since an imperfectly segmented shape could have a high number of interesting points which do not belong to the main object. With the higher-order constraints, those "fake" interesting points could be matched to the "real" one which influence the similarity value between shapes.

Figure 6.11: Sample images and their correlated shapes from the ETHZ [FJS09] database. The second row illustrates the ground truth shapes with manually added noise. The last row shows the segmented shapes using the shape generation results in Section 6.1.2. The third and the fourth rows show the results in the middle steps of shape generation.

Therefore, for the shape matching applications with imperfect segmentation, the high-order graph matching is not recommended. The second reason is due to the invariance properties of the proposed shape descriptor discussed in Section 4.2. In practice, this problem could be solved by some shape preprocessing methods [Yan+14a].

6.3.1.6 Computational Complexity

Here the computational complexity of the proposed interesting point generation and matching approaches is analysed. (1) For interesting point generation, the time complexity is in the order of $O(N'^2)$, where N' is the number of boundary points in the shape D. This is because the worst case of *Douglas-Peucker* [Ebi02] is $O(N'^2)$. Moreover, the computational complexity of FMM [HF07] is $O(N')$. For the remaining sequence generation and filtering tasks, they can be finished in $O(N')$ time. Fusing those tasks together, the overall complexity is $O(N'^2)$. (2) For point context generation, the time complexity is $O(mn)$, where m and n are the number of interesting points and contour sample points, respectively. This is because point context

Table 6.26: Experimental results of interesting point detection (the upper table) and shape retrieval (the lower table) in the case of imperfect segmentations. Here, M and S illustrate the ground truth shapes with manually added noise and the shapes using the generation method in Chapter 2, respectively.

	apples		bottles		giraffes		mugs		swans		**mean**	
	FP	FN	FP	FN	FP	FN	FP	FN	FP	FN	**FP**	**FN**
M	2.98	16.45	2.26	1.04	0.38	12.35	2.15	16.96	0.52	9.52	**1.66**	**11.26**
S	0.94	9.63	0.87	3.56	0.83	12.40	0.69	12.75	0.74	9.88	**0.81**	**9.64**

	apples	bottles	giraffes	mugs	swans	**mean**
M	83.81	79.99	68.01	71.09	54.88	**71.56**
S	70.50	60.81	62.07	55.12	48.63	**59.43**

features are generated by considering the distance and orientation relationships between each interesting point and contour sample points. In practice, since m is a value independent of contour sample points and $m \ll n$, the computational complexity of this method is determined by the contour sample points. Therefore, the complexity for feature generation is $O(n)$. (3) For interesting point matching, the computational complexity is analysed by different potentials. The Hungarian algorithm is used for singleton potential; as introduced in [Kuh55], it can solve the point matching task in $O(N^3)$ time. For the third-order potential, assuming m_1 and m_2 interesting points are selected from two shapes, there are $O(m_1^3 m_2^3)$ possible triplets, each represented by a high-order term in Equation (5.11). This enables the proposed method to significantly reduce the complexity without searching for all possible matching correspondences.

Based on the Kimia216 dataset and the experimental environment introduced above, the computation time is estimated and reported here. On average, the shape resolution in this dataset is 187×239. For interesting point generation, the mean time is 5.33 seconds on each shape. With interesting points, the mean time for point context generation in each shape is 0.01 seconds. Given two sets of interesting points, the mean matching time is 0.09 seconds. Please notice that the codes are not optimised, and its faster implementation is possible by optimising loops and programming, etc.

6.3.2 Object Retrieval using the Region-based Method

In this section, the proposed object representation method in Section 5.3.2 is firstly assessed formed on the singleton potential. Secondly, the usability of skeleton evolution is approved depending on the pairwise potential. Thirdly, the proposed matching algorithm with both potentials is evaluated. Lastly, the implementation and computational complexity of the proposed method is introduced and analysed.

The evaluations are built on a retrieval framework where shapes in the database are ranked based on their similarity to a query shape. To evaluate the retrieval performance, the following measure is used:

$$s = \frac{1}{100}\sum_{n=1}^{W} R_n(1 - \frac{n-1}{W}) \quad . \tag{6.1}$$

where W denotes the number of shapes which belong to the same class as the query shape. R_n denotes the number of retrieved shapes that are in the same class as the query in the top-ranked n shapes. The evaluation measure in Equation (6.1) is necessary to evaluate the retrieval performance accurately using both the number of correct matches and the ranking positions.

6.3.2.1 Evaluation of Hierarchical Skeleton-based Representation

This section aims to quantitatively assess the influence of skeleton pruning for shape matching as well as the effectiveness of hierarchical skeletons. For the first purpose, the PSSGM [BL08] method is employed for shape retrieval on the Kimia216 [SKK04] dataset. Table 6.27 depicts the matching performance of this method where skeletons are generated by a DCE-based approach with a fixed and manually tuned stop parameter k. Each shape is used as a query and the 12 most similar shapes among the whole dataset are retrieved. The final value in each position is counter values that are obtained by checking retrieval results using all the 216 shapes as queries. For example, the fourth position in the row of $k = 3$ shows that from 216 retrieval results in this position, 186 shapes are relevant to the query shapes. Scores in the last column are calculated with Equation (6.1). As shown in Table 6.27, results obtained by the fixed k are much worse than the result reported in [BLL07], as a consequence of the incompleteness of skeletons to represent shapes (e.g. body parts are missing in Figure 4.3). Like this, the matching performance heavily relies on the quality of skeletons and especially the stop parameter k.

For evaluating the performance of the proposed hierarchical skeleton representation, MPEG7 [LLE00] and Animal2000 [BLT09] databases are employed. The Animal2000 database has 2000 images where of 20 categories, each one consists of 100 images (Figure 6.12). Since shapes in Animal2000 are obtained from objects in real images, each class is characterised by a large intra-class variation of shapes. In particular, some important shape parts (e.g. legs) are missing, and some shapes have a noisy inside or outside (holes and patches).

Table 6.27: Experimental comparison of PSSGM [BL08] with different stop parameters k for contour partitioning with DCE. Results are summarised as the number of shapes from the same class among the first top 1-12 shapes. The last row represents the reported results in [BL08] where k is manually tuned for each shape.

k	1st	2nd	3rd	4th	5th	6th	7th	8th	9th	10th	11th	12th	Score
3	216	203	196	186	182	154	157	132	117	102	91	77	11.3567
4	216	205	190	182	171	157	148	136	126	117	104	102	11.3208
5	216	203	191	187	184	171	165	154	138	136	117	108	11.7908
6	216	205	188	191	178	181	173	148	148	122	115	98	11.8342
7	216	208	204	197	190	181	156	155	135	127	102	92	12.0067
8	216	205	198	195	197	193	189	176	155	126	120	93	12.3783
9	216	209	199	203	193	191	186	177	162	150	131	118	12.4567
10	**216**	**210**	**209**	**204**	**196**	**197**	**176**	**173**	**164**	**152**	**150**	**110**	**12.5817**
11	216	207	204	196	193	189	177	165	155	141	132	92	12.3550
12	216	209	202	197	188	185	162	163	157	140	137	93	12.2375
13	216	205	206	195	192	186	177	167	152	146	120	89	12.3100
14	216	208	204	199	198	184	185	166	153	158	139	103	12.4917
15	216	212	206	201	197	192	187	168	150	131	95	95	12.4608
[BL08]	216	216	215	216	213	210	210	207	205	191	177	160	13.6983

Table 6.28 and Table 6.29 show the performance comparison between single and hierarchical skeleton matching on MPEG7 and Animal2000, respectively. The performance in Table 6.28 is evaluated using bulls-eye scores, while the one in Table 6.29 is measured based on Equation (6.1). For single skeleton matching, the presented scores are obtained by different stop parameters k. [3,15] in both Table 6.28 and Table 6.29 illustrates the hierarchical levels $[T_{min}, T_{max}]$ for calculating the singleton potential (without fusing pairwise potentials). For a fair comparison, the same skeleton-based matching algorithm proposed by Bai [BL08] is used for single and hierarchical skeletons. Table 6.28 and Table 6.29 clearly indicate that without any human intervention (i.e. manual turning of a stop parameter), the proposed hierarchical skeletons perform better than the traditional single skeletons. In the following discussion, the best scores marked with bold font in Table 6.27, Table 6.28 and Table 6.29 will be used for the comparison of the proposed matching method.

Figure 6.12: Sample shapes from the Animal2000 [BLT09] database.

Table 6.28: Bulls-eye score on the MPEG7 dataset using single skeletons with different stop parameters k and hierarchical skeletons within the level [3,15]. Hierarchical skeleton matching outperforms the best single skeleton matching (marked with bold font).

k	3	4	5	6	7	8	9
bulls-eye	0.7125	0.7165	0.7325	0.7421	0.7405	0.7430	0.7516
k	10	**11**	12	13	14	15	[3,15]
bulls-eye	0.7451	**0.7547**	0.7452	0.7455	0.7464	0.7432	0.7884

6.3.2.2 Evaluation of Hierarchical Skeleton Evolutions

In this section, the aim is to visually prove that skeleton evolutions for the same category are more similar than those for different ones. To do so, nine objects are randomly selected for three different classes (three objects in each class) from the Kimia216 dataset [SKK04]. After that, the hierarchical skeleton of each object is generated with k from 3 to 28. As it is composed by single skeletons from coarse to fine levels, it can be treated as a sequence. The first class involves bird02, bird07 and bird10 sequences. The second class involves camel12, camel14 and camel17. The third class involves face01, face02, face03. Here, skeletons with $k = 3$ and $k = 28$ are the start and end frames of the sequence, respectively. In each frame in a sequence, the dissimilarity to the start frame is calculated using Equation (5.23). This dissimilarity calculation is applied to all nine sequences. It could happen that shapes in the same class would have different skeleton evolutions due to the scale difference. However, as illustrated in Equation (5.23) and Equation (5.24), both radii and skeleton lengths have been normalised, the proposed dissimilarity measure for skeleton evolution is scale invariant.

For each sequence, the dissimilarity between each frame and the start frame are plotted

Table 6.29: Scores are computed by Equation (6.1) on Animal2000 using the single skeletons with different stop parameters k and the proposed hierarchical skeletons within the level [3,15]. Hierarchical skeleton matching outperforms the best single skeleton matching (marked with bold font).

k	3	4	5	6	7	8	9
bulls-eye	231.55	225.98	234.49	228.09	230.73	225.93	232.61
k	10	11	**12**	13	14	15	[3,15]
bulls-eye	234.39	233.97	**241.33**	233.60	239.77	233.67	346.61

in Figure 6.13. Horizontal and vertical axes represent frame IDs (hierarchical levels) and dissimilarities to the start frame, respectively. For the sequences belonging to the same class, they are plotted with the same colour. In Figure 6.13 (a), the first (birds, red) and the second (camels, green) classes are printed. In Figure 6.13 (b), the first and the third (faces, green) classes are printed. For all nine objects, it can be clearly observed that 1) the monotonicity of the dissimilarity is evident throughout the sequence as well as 2) dissimilarity changes in the same class are more similar than the ones in different classes. This proves the capacity of skeleton evolutions for distinguishing topologically different shapes.

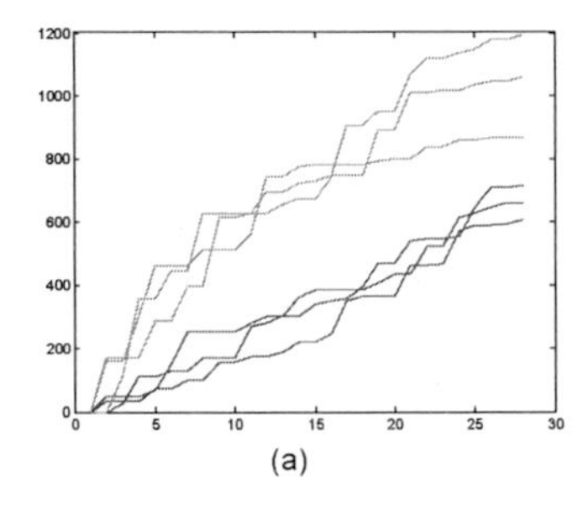

(a)

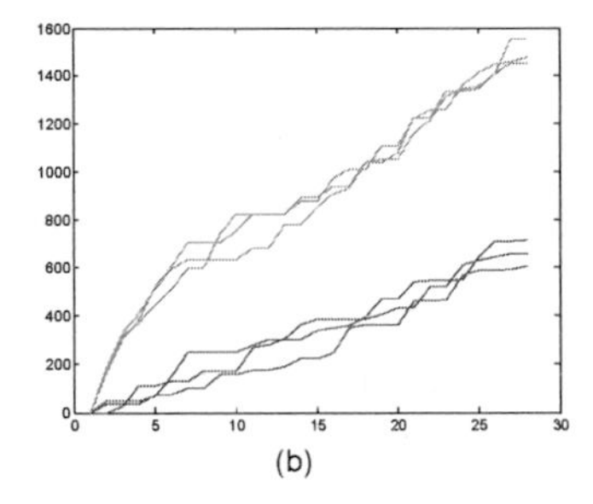

(b)

Figure 6.13: Dissimilarities of each frame to the first one in skeleton evolution sequences. (a) Dissimilarities in birds (red) and camels (green) sequences and (b) dissimilarities in birds (red) and faces (green) sequences.

Moreover, the performance improvement using skeleton evolution for object retrieval is also evaluated on the Kimia216 dataset. Based on Equation (6.1), matching scores of four methods are calculated along with the DCE steps. As shown in Figure 6.14, these methods are matching with both singleton and pairwise potentials (red bar), matching with singleton potential (blue bar), the globally optimum matching (green bar) and the single skeleton matching with PSSGM [BL08] (black bar). For example, on DCE step 5, the score of the red bar is calculated with the hierarchical skeletons from level 3 to 5 using both singleton and pairwise potentials while the blue bar uses only the singleton potential. The score of the green bar is calculated with the best matched skeletons among 3 to 5 levels while the black bar uses the skeletons only on level 5. Figure 6.14 shows that the global optimum matching obtains the lowest score among all DCE steps. The main reason for this is the overfitting between skeletons with little endpoints and skeletons with plentiful endpoints. For the PSSGM [BL08] method, the score is increasing from step 3 until step 10. After that, the score decreases. For the singleton potential and the proposed method, scores are gradually increasing. Even the singleton potential performs better than the PSSGM method in the majority of steps. With the benefits from both the singleton and pairwise potentials, the proposed method achieves the highest score. This validates the effectiveness of skeleton evolutions for object matching. In order to improve the matching performance, more levels can be used for constructing hierarchical skeletons. It will, however, incur more time for skeleton matching. Therefore, a proper level range of hierarchical skeletons should be clarified to balance the scores and the matching time. As illustrated in Figure 6.14, the matching performance of the proposed method is gradually increased from level 3 to 13 and then becomes stable from level 13 to 15. With this observation, 13 or its adjacent levels are selected as the general level range for all other experiments.

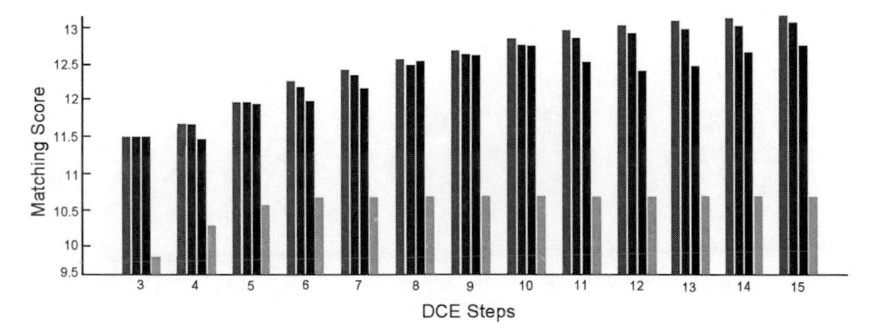

Figure 6.14: Score comparison between different matching methods: the proposed method (red), the singleton potential (blue), the global optimum matching method (green) and the PSSGM method (black). The horizontal axis represents the hierarchical skeleton levels and the vertical axis represents the retrieval score with Equation (6.1).

6.3.2.3 Hierarchical Skeleton-based Matching

In this section, the generality of the proposed method is evaluated using six datasets: Kimia216 [SKK04], Kimia99 [SKK04], Tari56 [AT05], Tetrapod120, MPEG7 [LLE00] and Animal2000 [BLT09].

Kimia216 Database: Table 6.30 shows the performance comparison between the proposed method, PSSGM (PS1) [BL08] with the fixed k which achieves the best score in Table 6.27, PSSGM (PS2) with manually tuned k for each shape, ID [LJ07] and SC [BMP02]. As shown in the upper block of Table 6.30, the proposed method performs better than PS1 with the best stop parameter and close to PS2 with manually tuned k for each shape.

Table 6.30: Experimental comparison of the proposed method to state-of-the-art methods on Kimia216 and Kimia99 datasets. PS1 denotes the retrieval results with fixed k which achieves the highest score with Equation (6.1). PS2 denotes the reported results in [BL08] with manually tuned k for each shape.

Kimia216	1st	2nd	3rd	4th	5th	6th	7th	8th	9th	10th
ID [LJ07]	216	198	189	176	167	156	136	130	122	118
SC [BMP02]	204	199	192	187	185	181	175	166	160	163
PS1	216	210	209	204	196	197	176	173	164	152
PS2 [BL08]	216	216	215	216	213	210	210	207	205	191
Proposed Method	**216**	**216**	**213**	**212**	**209**	**197**	**196**	**192**	**193**	**172**

Kimia99	1st	2nd	3rd	4th	5th	6th	7th	8th	9th	10th
ID [LJ07]	99	97	92	89	85	85	76	75	63	53
SC [BMP02]	99	97	91	88	84	83	76	76	68	62
PS1	99	97	97	97	96	92	93	81	71	68
PS2 [BL08]	99	99	99	99	96	97	95	93	89	73
Proposed Method	**99**	**99**	**99**	**96**	**94**	**95**	**91**	**89**	**85**	**77**

Kimia99 Database: With the Kimia99 dataset, shape retrieval results with the proposed method are compared to the other methods (best score from $k = 12$ with PSSGM [BL08]). As illustrated in Table 6.30, the proposed method significantly outperforms the other methods with a 1.85% improvement over PS1 (with Equation (6.1)) and close to the PS2 with manually tuned k for each shape.

Tari56 Database: The Tari56 database is used for testing the performance on non-rigid objects. It includes 14 classes of articulated shapes with 4 shapes in each class (Figure 6.15).

Figure 6.15: Full shapes from the Tari56 [AT05] database.

Although this database was introduced in [AT05], no result on the whole database is presented. Moreover, there is no detailed explanation of retrieval results on this dataset using ID [LJ07], PSSGM [BL08] and SC [BMP02] methods. In this part, those methods are all applied to the Tari56 database and their retrieval performances are compared. Since there are some parameters involved by ID and SC methods, they are assigned by the optimisation method introduced in Section 5.2.2.3. The retrieval results are reported in Table 6.31 (the result of PSSGM is obtained by the best stop parameter $k = 10$). It can be observed that the proposed method performs well (1.96% better than the PSSGM method) in the presence of non-rigid deformations. This is because hierarchical skeletons can capture non-rigid features of shapes by multiple skeletons, which have a higher discrimination power than single skeletons used in other methods.

Table 6.31: Results comparison on the Tari56 dataset.

Tari56	1st	2nd	3rd	4th
ID [LJ07]	56	46	37	28
SC [BMP02]	52	17	10	10
PSSGM [BL08]	56	49	44	40
Proposed Method	**56**	**51**	**50**	**33**

Tetrapod120 Database: The aim of using the Tetrapod120 dataset is to evaluate the ability of matching methods for fine-grained shapes where some species are really difficult to distinguish (e.g. horses and dogs). The main skeleton structures among all these objects are quite similar. Therefore, in order to improve the accuracy for object matching, fine-grained

skeleton branches should be considered. As shown in Table 6.32, the proposed method achieved the best results among all the other methods (8.96% better than the best score from ID [LJ07] with Equation (6.1)). Although only 15 DCE levels are used for the experiment, there are still some levels unused but with essential shape features. Therefore, the overall performance could be improved by integrating edit distance [TH04] for measuring fine-grained deformations using these unused hierarchical levels.

Table 6.32: Experimental comparison on the Tetrapod120 dataset.

Tetrapod	1st	2nd	3rd	4th	5th	6th	7th	8th	9th	10th
ID [LJ07]	120	118	106	101	90	83	77	69	70	56
SC [BMP02]	100	80	70	53	53	51	40	28	27	27
PSSGM [BL08]	120	109	101	98	81	78	68	66	65	59
Proposed Method	**120**	**118**	**106**	**100**	**95**	**90**	**84**	**71**	**83**	**81**
	11th	12th	13th	14th	15th	16th	17th	18th	19th	20th
ID [LJ07]	57	45	38	29	41	35	26	27	30	21
SC [BMP02]	29	27	25	32	32	23	31	26	20	28
PSSGM [BL08]	59	49	50	42	43	35	39	31	33	36
Proposed Method	**68**	**73**	**67**	**77**	**68**	**67**	**60**	**51**	**56**	**43**

MPEG7 Database: For comparison, Table 6.33 lists several reported results and the results by the proposed method on the MPEG7 dataset. Similar to the experiment in Table 6.25, those methods are clustered into two groups: pairwise matching and context-based matching. The proposed Hierarchical Skeleton with both singleton and pairwise potentials achieves a 81.62% bulls-eye score that is significantly better than those of traditional skeleton-based methods [XHS08; BLL07]. This is because a single skeleton has limited ability to capture geometric properties at different levels of resolution. However, this approach performs not as well as Shape Tree [FS07] and Height Functions [Wan+12a], etc. The main reason for this is the maximal DCE levels for the experiment being limited to 15 while the performance can be improved by using more DCE levels.

In the group of context-based methods, similar to Table 6.25, the MG method [KDB10] is employed based on the similarity scores between all hierarchical skeletons. The proposed method, which achieves 99.21% bulls-eye score, outperforms most state-of-the-art methods.

Animal2000 Database: As shown in Table 6.34, the proposed method performs better than the SC [BMP02] and PSSGM [BLL07] methods. However, there is still some space to improve

Table 6.33: Bulls-eye score on the MPEG7 Dataset.

Pairwise Matching	Score	Context-based	Score
Shape Contexts [BMP02]	76.51%	INSC + CDM [Jeg+10]	88.30%
Skeletal Context [XHS08]	79.92%	IDSC + LP [Bai+10]	91.00%
Optimized CSS [MB03]	81.12%	SC + LP [Bai+10]	92.91%
Multiscale Rep. [AO04]	84.93%	IDSC + LCDP [YKTL09]	93.32%
Shape L'AneRouge [PRH08]	85.25%	SC + GM + Meta [EKG10]	92.51%
Fixed Cor. [Sup06]	85.40%	IDSC + MG [KDB10]	93.40%
Inner Distance [LJ07]	85.40%	IDSC + PSSGM + LDCP [Tem+10]	95.60%
Symbolic Rep. [DT08]	85.92%	ASC + LDCP [LYL10]	95.96%
Hier.Procrustes [MV06]	86.35%	HF + LCDP [Wan+12a]	96.45%
Triangle Area [AKF08]	87.23%	SC + IDSC + Co-T [Bai+12]	97.72%
Shape Tree [FS07]	87.70%	SC + DDGM + Co-T [Bai+12]	97.45%
Height Functions [Wan+12a]	89.66%	AIR [GTC10]	93.67%
IP [Yan+15a]	80.28%	IP+HG [Yan+15a]	96.43%
PSSGM [BLL07]	75.16%	ASC + TN + TPG [YPL13]	96.47%
Proposed Method	**81.62%**	**Proposed Method + MG**	**99.21%**

this method. In particular, the shape noise has a bad influence on the skeleton generation and pruning. This is why the retrieval score of ID [LJ07], which is robust to shape noise, is higher than the score of skeleton-based methods. In addition, some shapes in the same class are significantly different (Figure 6.12). Therefore, it is not possible to group them into the same class only using shape distance methods. In Section 8.2, a further discussion is addressed that this dataset will be used for evaluating the tasks with a supervised learning process, e.g. shape-based object classification using hierarchical skeletons.

Table 6.34: Retrieval score comparison on the Animal2000 dataset.

Method	ID [LJ07]	SC [BMP02]	PSSGM [BLL07]	**Proposed Method**
Score	452.66	193.79	241.33	**348.73**

6.3.2.4 Computational Complexity

All the implementation steps are briefly described: first, an initial skeleton is computed with the method in [CLS03]. After that, a hierarchical skeleton is generated using the skeleton pruning method in Section Section 4.3. Then, the hierarchical skeletons are matched by two potentials. For the singleton potential, the skeleton matching method is employed in which Dijkstra's shortest path algorithm [Val+13] is employed to build the skeleton graph [BL08]. For the pairwise potential, skeleton lengths and skeleton point radius are calculated to obtain the distance between two skeletons on different hierarchical levels. Finally, the total costs between hierarchical skeletons are computed with the proposed method in Section 5.3.2.

Now, the computational complexity of the above steps is analysed: (1) For initial skeleton generation, the time complexity is in the order of $O(8N'')$, where N'' is the number of points in the planar shape D. This is because in [CLS03], to determine whether a pixel point in D is a skeleton point, the corresponding nearest contour point for each of the 8 neighbouring points is determined. After removing the constant value in $O(8N'')$, the time complexity becomes $O(N'')$. (2) For hierarchical skeleton generation, the time complexity is $O(v' \log v')$, where v' is the number of the vertices in the original polygon P. This is because, as introduced in [BLL07], the skeleton pruning based on DCE has a complexity of $O(v' \log v')$. (3) For hierarchical skeleton matching, the computational complexity is analysed by different potentials. Assuming that k_1 and k_2 are the number of all nodes (skeleton endpoints and junction points) in two single skeletons, the time complexity for computing the singleton potential is $O(k_1^2 k_2^2)$ [BL08]. As there are $T_{max} - T_{min} + 1$ hierarchical levels employed for the singleton matching, the total complexity is $O(T_{max} - T_{min} + 1) \times O(k_1^2 k_2^2)$. Since $T_{max} - T_{min} + 1$ is constant, the total complexity for the singleton potential is $O(k_1^2 k_2^2)$. For the pairwise potential, the time for computing DT is $O(N'')$ by the approach proposed in [MQR03]. With DT, skeleton evolution is computed by considering all possible pairs of skeletons on two different levels, the time complexity is $O((T_{max} - T_{min} + 1)(T_{max} - T_{min})/2)$. Considering these two parts, the pairwise potential runs in $O(N'') + O((T_{max} - T_{min} + 1)(T_{max} - T_{min})/2)$ time. Since $(T_{max} - T_{min} + 1)$ and $(T_{max} - T_{min})$ are constant and small, the total computational complexity of the pairwise potential is $O(N'')$.

Here the computation time is reported based on the Kimia216 dataset and the experimental environment introduced above. On average, the shape resolution in this dataset is 187×239. For initial skeleton generation, the mean time is 0.07 seconds on each shape. With an initial skeleton, the mean time for hierarchical skeleton generation is 0.97 seconds. Given two hierarchical skeletons with levels $[3, 15]$, the mean matching time is 11.29 seconds. Please notice that the employed codes are not optimised, and its faster implementation is possible by optimising loops, settings and programming language, etc. Thus, there are still plenty of opportunities to reduce the running time.

6.4 2D Object Retrieval using Integrated Features

In this section, three integrated shape descriptors in Section 5.4 and their correlated matching algorithms are assessed. The experiments in the three subsections are all performed on a laptop with Intel Core i7 2.2GHz CPU, 8.00GB memory and 64-bit Windows 8.1 OS. All methods in the experiments are implemented in Matlab R2015a.

6.4.1 Object Retrieval with Shape Context and Boundary Segments

The integrated shape descriptor and its matching algorithm in Section 5.4.1 are evaluated on two shape databases provided by Kimia216 [SKK04] and MPEG400. The proposed method is compared to the PSSGM [Hed+13] (PSSGM in Table 6.35) and the boundary segments-based matching method in Section 6.2.1 (Boundary Segments in Table 6.35). To compare the interesting point extraction technique against existing methods, the proposed method with interesting points gained from the DCE approach [BLL07] is also tested (DCE-based in Table 6.35). Similar to previous experiments, for each of the shapes used as a query, the retrieved results are checked whether they are correct, i.e. belonging to the same class as the query. Results achieved for the Kimia216 and the MPEG400 datasets can be found in Table 6.35, whereas in [Hed+13] the PSSGM algorithm had been re-evaluated without any preliminary assumptions regarding the object skeletonisation. For these two databases, it is important to observe that the fused shape context and boundary segment descriptor outperforms the skeleton-based method [Hed+13] and the boundary segment-based method while the complexity of feature generation and matching has been reduced.

6.4.2 Object Retrieval with Skeleton and Boundary Segments

To evaluate the proposed methodology in Section 5.4.2, the experiments in this part are performed in an object retrieval scenario using two different datasets: Kimia216 [SKK04] and MPEG400. Similar to the experiments in Section 6.4.1, for Kimia216 and MPEG400, the proposed algorithm is run in two configuration modes and compared to the PSSGM [BL08] method. In the first configuration, only boundary segments have been used for object description, whereby the open curve descriptor introduced in Section 3.2 together with the dissimilarity measure proposed in Section 5.2.1 are used. In the second mode, properties of both, boundary segments and skeletons, are extracted for object representation. In both cases, skeletons computed according to [BLL07] have been used for partitioning a shape boundary into multiple boundary segments. In order to enable quantitative comparison, the experimental convention in previous experiments is kept and the 10 best matches are considered for each query. Results achieved for the Kimia216 and the MPEG400 datasets can

Table 6.35: Experimental comparison of the proposed methodology to the most related algorithm using the MPEG400 and Kimia216 datasets. Results are summarised as the number of shapes from the same class to the query among the first top 1-10 shapes.

MPEG400	1st	2nd	3rd	4th	5th	6th	7th	8th	9th	10th
PSSGM [Hed+13]	380	371	361	351	344	339	332	320	330	309
Boundary Segments	375	348	333	325	317	311	300	295	276	275
DCE [LL99]-based	375	368	346	337	338	323	308	297	286	276
Proposed Method	**389**	**374**	**368**	**368**	**358**	**347**	**344**	**339**	**346**	**330**

Kimia216	1st	2nd	3rd	4th	5th	6th	7th	8th	9th	10th
PSSGM [Hed+13]	205	208	202	199	200	192	184	167	161	130
Boundary Segments	216	215	206	204	200	186	172	163	130	124
DCE [LL99]-based	216	204	197	185	175	162	154	154	142	131
Proposed Method	**216**	**214**	**207**	**204**	**201**	**204**	**191**	**188**	**192**	**185**

be found in Table 6.36, whereas in [Hed+13] the PSSGM algorithm has been re-evaluated without any preliminary assumptions regarding the object skeletonisation. Please be noted that the boundary segments in Table 6.36 are different from the one in Table 6.35 since the boundary segments in Table 6.36 are generated by the skeleton endpoints rather than the DCE method in Table 6.35.

Table 6.36: Experimental comparison of the proposed methodology to the related algorithms using the Kimia216 and the MPEG400 datasets.

MPEG400	1st	2nd	3rd	4th	5th	6th	7th	8th	9th	10th
PSSGM [Hed+13]	380	371	361	351	344	339	332	320	330	309
Boundary Segments	375	348	333	325	317	311	300	295	276	275
Proposed Method	**383**	**373**	**364**	**356**	**349**	**343**	**336**	**320**	**330**	**312**

Kimia216	1st	2nd	3rd	4th	5th	6th	7th	8th	9th	10th
PSSGM [Hed+13]	205	208	202	199	200	192	184	167	161	130
Boundary Segments	216	215	206	204	200	186	172	163	130	124
Proposed Method	**216**	**216**	**214**	**213**	**213**	**211**	**204**	**193**	**184**	**175**

It can be clearly observed that the combination of shape boundary segments and skeleton properties significantly improves the effectivity of the shape retrieval methodology. Moreover, since the time complexity for boundary segment representation and matching is rather low, the overall computational complexity of the integrated methods is close to the time complexity of PSSGM introduced in [BL08].

Table 6.37: Experimental comparison of the SC descriptor to its fused descriptor in Section 5.4.3 using the Kimia216 and the MPEG400 datasets.

MPEG400	1st	2nd	3rd	4th	5th	6th	7th	8th	9th	10th
SC [BMP02]	400	370	343	310	302	277	272	265	264	239
Proposed Method	**400**	**389**	**374**	**368**	**368**	**358**	**347**	**344**	**339**	**346**

Kimia216	1st	2nd	3rd	4th	5th	6th	7th	8th	9th	10th
SC [BMP02]	216	212	201	187	186	175	171	164	162	146
Proposed Method	**216**	**214**	**207**	**204**	**201**	**204**	**191**	**188**	**192**	**185**

6.4.3 Object Retrieval with Shape Context and Bounding Boxes

In order to proof the applicability of the proposed approach in Section 5.4.3, a set of evaluations are conducted on Kimia216 [SKK04] and MPEG400 datasets. The first experiment is performed only using the SC descriptor with the same experimental set up in [BMP02]. After that, the integrated descriptor is employed with the fusing weight introduced in Equation (5.27). Results achieved for both datasets can be found in Table 6.37 in which the 10 best matches are considered for each query and the results are summarised as the number of shapes from the same class as the queries. Table 6.37 shows that the fused descriptor achieves significant progress on both databases.

Chapter 7

Applications

In this chapter, based on the proposed coarse- and fine-grained shape descriptors and their matching algorithms, two shape-based applications are introduced to prove the applicability of the proposed methods to real-world problems.

7.1 Shape-based Applications

Shape-based methods have been widely used for various applications in computer vision, e.g. character recognition [YT07], biomedical image analysis [Dar+06; Wan+11; Bau+15], hand gesture recognition [Ren+13], object tracking [Kha+15], robot navigation [WL04], human gait recognition [CT13; Yan+16b], etc. For character recognition, as shown in Figure 7.1, a skeleton model is normally employed for capturing the geometrical and topological features of given characters. After that, skeleton graph matching methods [BL08; Fei16; SK05; HPF14] can be used for character recognition.

Figure 7.1: Skeletons (b) of two Chinese characters (a).

For biomedical image analysis, Daras *et al.* [Dar+06] proposed a method for classifying proteins (Figure 7.2) based on the comparison between their three-dimensional shape structures. With the portioned FSSP protein database, Wang *et al.* [Wan+11] assessed the efficiency of their proposed shape similarity learning method in a protein retrieval framework.

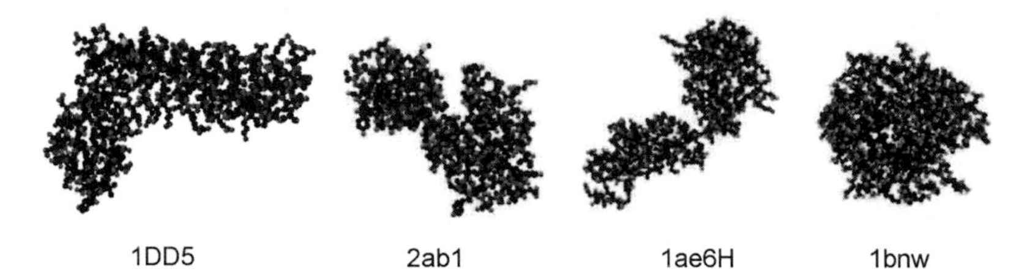

Figure 7.2: Sample proteins in the FSSP protein dataset illustrated in [Dar+06].

In [Ren+13], the part-based shape representation and its distance methods are introduced for hand gesture recognition. This method can handle the noisy hand shapes obtained by a Kinect camera [Sho+13]. In addition to hand gesture recognition, Choudhury *et al.* [CT13] proposed a shape-based method for human gait recognition (Figure 7.3), where the shape changes are analysed using a Fourier-based descriptor on shape contours.

Figure 7.3: A sample shape sequence for human gait recognition.

Based on a region of arbitrary shape, Khan *et al.* [Kha+15] proposed a shape-tailored local descriptor for textured object tracking which also achieved a promising performance on object segmentation. For robot navigation, Wolter and Latecki [WL04] presented a geometric model for robot mapping based on shape. Specifically, they proposed a structural shape representation of the robot's surrounding that grants access to metric information as needed in path planning. The structural shape representation is built on a boundary-based approach by considering the discrete structure provided by sensors, using polygonal lines to represent the boundaries of obstacles.

Different from the above applications, in this chapter, two shape-based applications are reported by bridging the domain of biology and audio signal processing with computer vision. These applications are applied based on the proposed coarse- and fine-grained shape descriptors.

7.2 Application using Coarse-grained Features

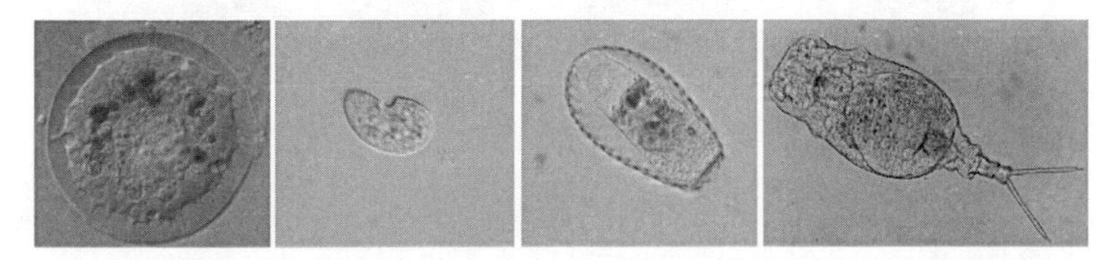

Figure 7.4: Sample images of EM.

Based on the proposed coarse-grained shape descriptor in Section 3.3, a shape-based application for Environmental Microorganism (EM) classification is designed. EMs (Figure 7.4) are microscopic beings living in natural (rivers, seas, forests, mountains, etc.) and artificial (fields, gardens, fish ponds, aeration tanks, etc.) surroundings. Their classification is a very important indicator for biological treatment processes and environmental quality evaluations. Unfortunately, it is very difficult to distinguish thousands of EMs from each other. Traditionally, they are recognised manually in environmental laboratories either by observing their shapes under a microscope (the morphological method [MG10]) or using molecular biology techniques. Although the morphological approach has the lowest cost in terms of time and expense, even very experienced operators are unable to distinguish thousands of EMs without referring to literature. The molecular technique distinguishes EMs by Deoxyribonucleic Acid (DNA) or Ribonucleic Acid (RNA) [GSL91; Ber+95] and is very accurate, even so it is slow and expensive.

In order to overcome the problems of these two methods, a practical and efficient system is developed in which microscopic images are automatically analysed to perform the EM classification. The proposed methodology simulates the morphological approach exploiting and modelling shape properties of EMs. First, the system conducts image segmentation to obtain EM shapes. Then, based on the shape descriptor in Section 3.3, features which characterise the shape of each EM are extracted from these segmented images. Afterwards, the class of the EM in an image is determined by a classifier based on its shape features. Finally, the classification results are returned to the users.

EM Shape Generation: There exist different segmentation methods based either on pixel intensity levels or on image context. In this application, a semi-automatic EM segmentation method introduced in [Li+13a] is employed for EM shape generation. This method is based on the Sobel edge detector [Can86] which has low sensitive to noise and easy to control. Samples of generated EM shapes are illustrated in Figure 7.5.

EM Shape Representation and Classification For EM shape representation, the proposed

Figure 7.5: Sample EM shapes based on the segmentation method in [Li+13a].

shape features in Section 3.3 are employed. However, the last feature, shape skeleton length, is not used for EM shape representation. The main reason is that some EM shapes are close to a circle (e.g. the first shape in Figure 7.5) in which the skeleton is just one point in a circle center. In such case, the skeleton length feature cannot be used to distinguish circle-similar shapes. Therefore, only the top nine features in Equation (3.7) are used for shape representation.

For shape classification, a multi-class SVM is applied and optimised by the methods introduced in Section 5.2.2.2 and 5.2.2.3, respectively. Specifically, it uses a two-class SVM for each pair from a set of all considered classes $\{\omega_1, \omega_2, \cdots, \omega_N\}$. Thus, if there are N classes in total, $N(N-1)/2$ two-class classifiers have to be used. Initially, a sample pattern (query pattern) is classified using all these two-class SVMs. The final classification result is determined by counting to which class the sample pattern has been assigned most frequently.
Evaluation: For evaluation, a real dataset acquired in environmental laboratories of the University of Science and Technology Beijing is employed. It contains ten classes of EMs $\{\omega_1, \omega_2, \cdots, \omega_{10}\}$. Each class is represented by twenty microscopic images. These images are segmented semi-automatically with the method introduced in [Li+13a]. Examples of EM images can be seen in Figure 7.6. The manually segmented images (ground truth) are also illustrated in this figure.

For supervised training, 10 manually segmented samples are randomly selected from each class. For testing the 10 remaining manually segmented images and all 20 semi-automatically segmented samples from each class are used separately. In such a case, two groups of classification results, manually and semi-automatically, are obtained. In order to increase statistical relevance, the selection process is repeated 10 times which leads to 10 different training datasets. The test datasets are changed only for experiments with manually segmented images. Experiments are performed for all these datasets and mean recognition rates are considered for evaluation.

As illustrated in Table 7.1, the proposed shape descriptor introduced in Section 3.3 possesses significantly better discriminative properties in comparison to the work in [Li+13b; Li+13a] for both the manually segmented and the semi-automatically segmented EM shapes. While for manually segmented shapes the overall classification rate increased from 89.7%

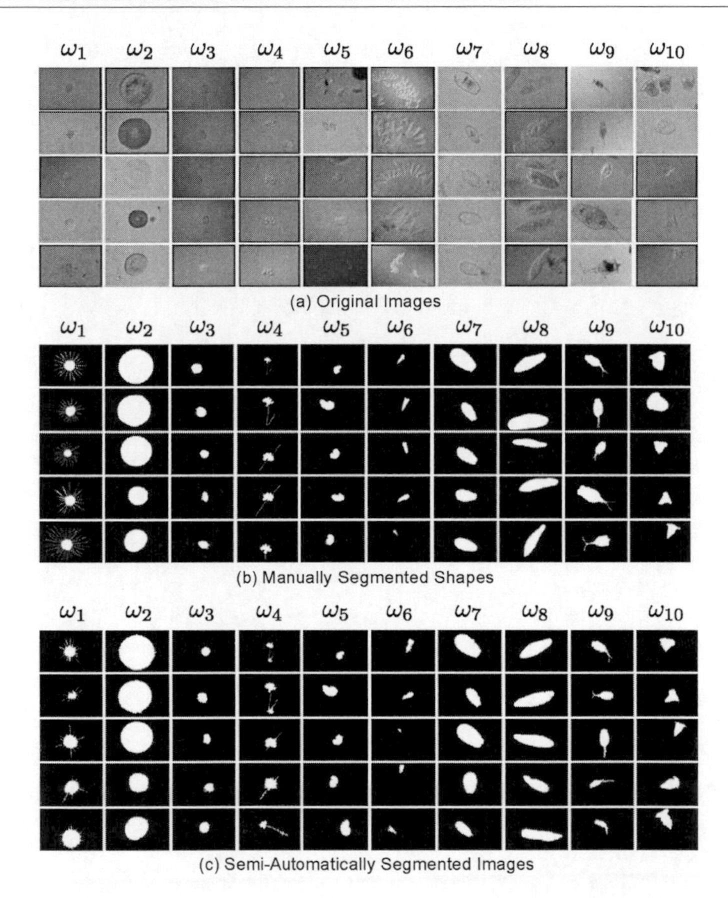

Figure 7.6: Examples of original, manually segmented, and semi-automatically segmented EM images of all 10 classes.

to 92.5%, the benefits of the proposed shape descriptor are even more significant for semi-automatically segmented images, namely from 66% to 79.5%, respectively. This result is very promising in real applications in which the microorganisms are not expected to be fully-automatically recognised by a computer-based system. However, as can be seen in Table 7.1, there are errors in approximately 20% of the classification rates for semi-automatically segmented shapes with the proposed feature space. This is due to the misclassification for class ω_4 and ω_6 in which the shapes are quite unstable (Figure 7.6) due to the non-rigid EM

Table 7.1: Classification rates for manually (first row) and semi-automatically (second row) segmented images achieved using the techniques introduced in [Li+13b; Li+13a] (left) and the shape descriptor proposed in Section 3.3 (right). M illustrates the mean value.

Manually Segmented Shapes [%]

Results in [Li+13b]

	ω_1	ω_2	ω_3	ω_4	ω_5	ω_6	ω_7	ω_8	ω_9	ω_{10}	
ω_1	**99**	0	0	1	0	0	0	0	0	0	99
ω_2	6	**90**	0	2	0	0	0	0	0	2	90
ω_3	0	0	**84**	6	2	2	0	0	0	6	84
ω_4	2	0	5	**93**	0	0	0	0	0	0	93
ω_5	0	0	4	0	**93**	3	0	0	0	0	93
ω_6	0	0	0	4	4	**92**	0	0	0	0	92
ω_7	1	0	0	2	0	0	**89**	0	0	8	89
ω_8	7	0	0	1	0	0	0	**89**	3	0	89
ω_9	2	0	0	4	0	0	0	2	**90**	2	90
ω_{10}	3	1	7	0	4	0	7	0	0	**78**	78
M											**89.7%**

Proposed Method

	ω_1	ω_2	ω_3	ω_4	ω_5	ω_6	ω_7	ω_8	ω_9	ω_{10}	%
ω_1	**93**	0	0	0	0	0	7	0	0	0	93
ω_2	0	**100**	0	0	0	0	0	0	0	0	100
ω_3	1	0	**77**	0	18	0	0	0	0	4	77
ω_4	0	0	5	**84**	0	0	0	2	9	0	84
ω_5	0	0	7	0	**93**	0	0	0	0	0	93
ω_6	0	0	0	0	5	**95**	0	0	0	0	95
ω_7	0	0	0	0	1	0	**99**	0	0	0	99
ω_8	0	0	0	0	0	0	0	**100**	0	0	100
ω_9	0	0	0	0	0	0	0	0	**100**	0	100
ω_{10}	0	0	0	0	1	0	15	0	0	**84**	84
M											**92.5%**

Semi-Automatically Segmented Shapes [%]

Results in [Li+13b; Li+13a]

	ω_1	ω_2	ω_3	ω_4	ω_5	ω_6	ω_7	ω_8	ω_9	ω_{10}	
ω_1	**70**	0	0	30	0	0	0	0	0	0	70
ω_2	10	**90**	0	0	0	0	0	0	0	0	90
ω_3	0	0	**100**	0	0	0	0	0	0	0	100
ω_4	0	0	15	**80**	0	5	0	0	0	0	80
ω_5	0	0	20	0	**55**	0	0	0	0	25	55
ω_6	0	0	0	55	0	**40**	0	0	5	0	40
ω_7	5	0	0	0	0	0	**5**	15	25	50	5
ω_8	15	0	0	0	0	0	0	**75**	10	0	75
ω_9	0	0	0	5	0	0	0	0	**95**	0	95
ω_{10}	20	0	0	0	15	0	0	0	15	**50**	50
M											**66.0%**

Proposed Method

	ω_1	ω_2	ω_3	ω_4	ω_5	ω_6	ω_7	ω_8	ω_9	ω_{10}	
ω_1	**50**	5	25	5	0	0	0	0	5	10	50
ω_2	0	**100**	0	0	0	0	0	0	0	0	100
ω_3	0	0	**85**	0	15	0	0	0	0	0	85
ω_4	0	0	25	**40**	5	15	0	5	5	5	40
ω_5	0	0	15	0	**85**	0	0	0	0	0	85
ω_6	0	0	5	0	40	**55**	0	0	0	0	55
ω_7	0	0	0	0	5	0	**95**	0	0	0	95
ω_8	0	0	0	0	0	0	0	**100**	0	0	100
ω_9	0	0	0	0	0	0	0	0	**100**	0	100
ω_{10}	0	0	5	0	0	0	10	0	0	**85**	85
M											**79.5%**

deformations. Moreover, comparing results of ω_6 between two segmentation methods with the proposed descriptor, there are only 5 misclassifications in manually segmented shapes, but 54 misclassifications in semi-automatically shapes. This observation reveals that there is still a lot of room for improving the employed EM segmentation method.

7.3 Application using Fine-grained Features

This part presents an application using skeleton model for audio signal analysis. The audio signal is represented in the time domain as a waveform, i.e. as a shape of the change of the air pressure as recorded from the microphone or generated synthetically. The waveform has a Temporal Fine Structure (TFS) changing rapidly and a slower varying part called the envelope [Moo08] (Figure 7.7). The envelope is often regarded as a signal modulating the TFS component (carrier).

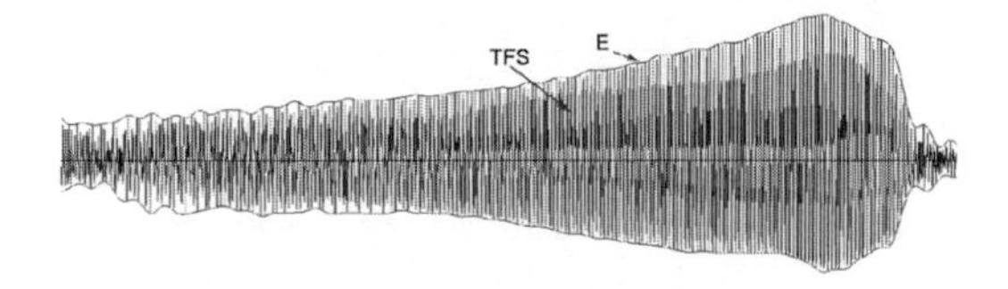

Figure 7.7: Temporal fine structure (TFS) and envelope (E).

In many applications, the information about the signal temporal envelope is indispensable, e.g. in the analysis of the timbre of musical instruments, the playability of string instruments, audio transients, speech intelligibility, audio coding and audio classification [AJ11; VS13]. In those applications, the envelope of a signal is broadly defined as slow changes in time of the signal, whereas the temporal fine structure is associated with its fast changes in time which could be treated as the carrier wave(s) of the signal [SDD12] (e.g. a violin vibrato sound in Figure 7.8). Many methods have been proposed to determine an envelope of a signal. Good reviews of existing methods have been given in [CR11].

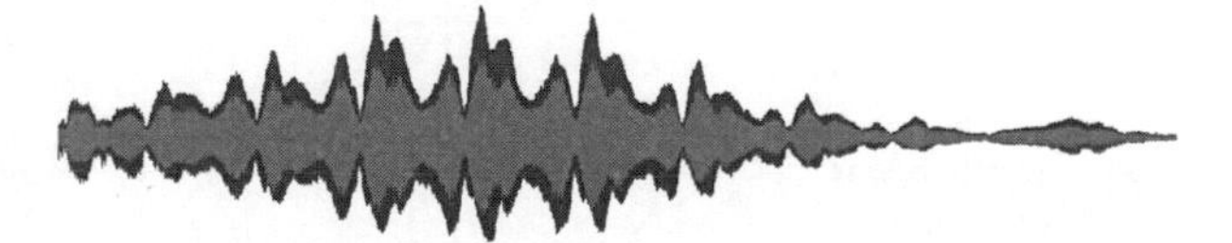

Figure 7.8: Envelope of violin vibrato sound.

However, there is neither an explicit mathematical definition, nor clear physical meaning of envelope [Qin+13]. To address this problem, several approaches derive from methods of time series analysis [KK03] to autoregressive modelling [AE07]. Another possible solution is to employ the computer vision methods to some classes of audio signals. Though the shape-based method for audio analysis has been proposed in [SMW08] for a specific signal (heart sounds), there is room for further search for a generic audio envelope shape descriptor.

In this part, a completely new approach to the problem of sound temporal envelope representation is proposed, using the method developed in Section 4.3. Specifically, in order to efficiently represent an audio envelope, the skeleton-based model, namely audio skeleton, is proposed as the descriptor for representing audio envelop features. This is because the skeleton model integrates both geometrical and topological features of the an audio envelop. Moreover, an efficient and fast matching algorithm is proposed based on the context features of audio skeleton endpoints. For evaluation, the proposed methods are applied on the the audio envelope for violin sound.

Audio Shape Generation: For a given audio excerpt, the audio envelope is generated based on its waveform, which is a time domain representation of a signal showing how its instantaneous amplitude varies over time. The amplitude envelope in the time domain is the line representing the evolution of the maximum amplitude of a waveform over time. It is a smooth curve outlining waveform extremes. For audio shape generation, a modified True Amplitude Envelope (TAE) approach [YGL15] is applied on the rectified waveforms. Specifically, to make the signal more compact in horizontal, a coefficient η' (from η' samples the maximum amplitude is retained) is necessary to downscale (zooming out) the signal. A cut-off frequency κ controls the smoothness of the envelope. In order to outline the extremes of the waveform, an iterative procedure is independently performed in which the amplitude peaks of the signal positive and negative parts are selected and returned. In this case, the bigger the number of iterations, the better the envelope matches the amplitude peaks of the signal. Figure 7.9 (a) represents a set of audio amplitude envelopes of a fragment of the violin music calculated for 50 iterations ($\eta' = 50$) at a fixed smoothness power κ, while Figure 7.9 (b) presents the signal and its envelope.

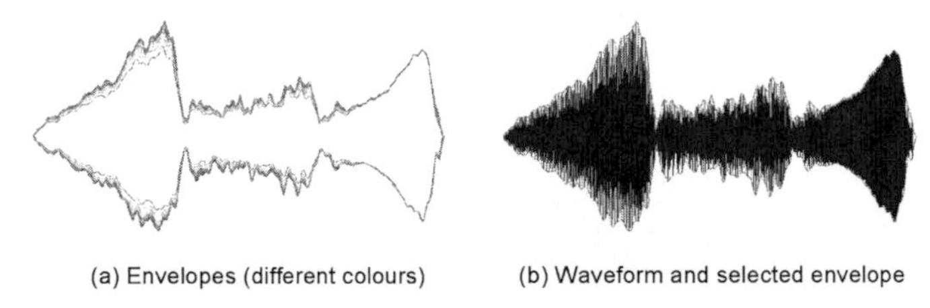

(a) Envelopes (different colours) (b) Waveform and selected envelope

Figure 7.9: Iterative generation and selection of an audio envelope.

Audio Shape Representation: Based on a selected audio shape, the skeletonisation method

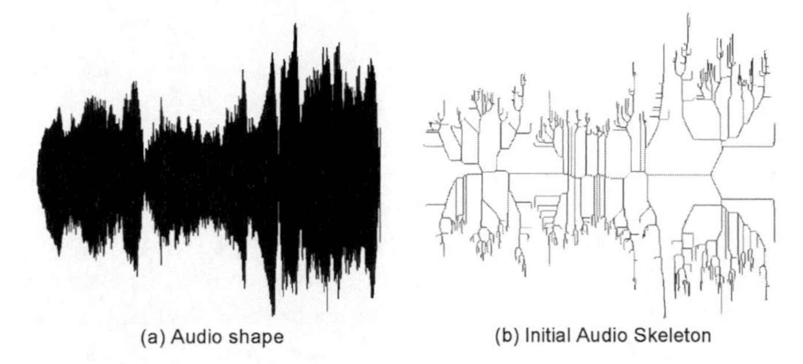

Figure 7.10: The skeleton in (b) has many redundant branches. This is mainly because of the boundary noise of shape (a).

introduced in Section 4.1 is used for initial audio skeleton generation. However, as illustrated in Figure 7.10, even though an audio envelope is smoothed before the skeletonisation process, there still exist some small skeleton branches. Figure 7.11 (a) illustrates an example of an audio skeleton where the regions with small skeleton branches are marked with blue circles. Therefore, a skeleton pruning process is required.

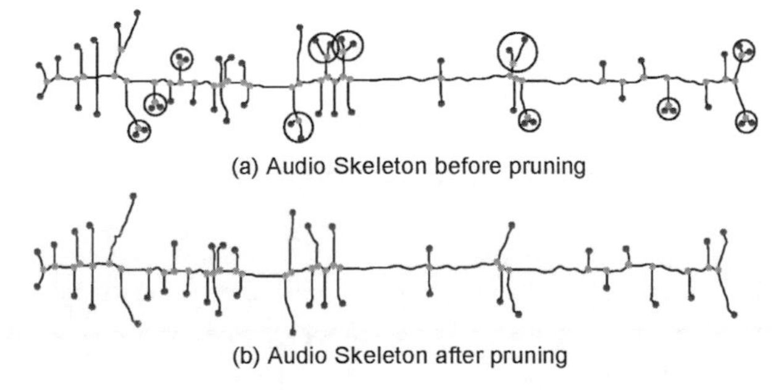

Figure 7.11: An audio skeleton before and after pruning.

Benefiting from the special topological structure of the proposed audio skeleton, most junction points (the green points in Figure 7.11 (a)) are located in the middle area of skeleton graphs and the junction points for redundant skeleton branches are distant from middle area. With this observation, for a skeleton graph S', its endpoints and junction points are denoted as $\mathbf{E}$ and $\mathbf{J}$. The borders of the middle area are calculated by:

$$\begin{aligned} \mathcal{B}_l &= \min(\mathbf{J}_h), & \mathcal{B}_r &= \max(\mathbf{J}_h) \\ \mathcal{B}_t &= \text{mean}(\mathbf{J}_v) - \delta, & \mathcal{B}_b &= \text{mean}(\mathbf{J}_v) + \delta \end{aligned} \quad . \tag{7.1}$$

where $\mathcal{B}_l$ and $\mathcal{B}_r$ are the left and right borders, $\mathcal{B}_t$ and $\mathcal{B}_b$ are the top and bottom borders, respectively. $\mathbf{J}_h$ and $\mathbf{J}_v$ denote the horizontal and vertical locations of all junction points, respectively. δ is the parameter for controlling the range of vertical borders. It should be determined by the scale of skeleton and the distribution of all junction points. More specifically, the δ is chosen if the ratio between the number of renewed junction points in the range of vertical borders and the increment of δ is at maximum. This strategy will ensure most junction points are located in a proper range of the vertical border.

For the middle area, the junction points $\mathbf{J}$ can be divided into two groups: inside the middle region ($\mathbf{J}'$) and outside the middle region ($\mathbf{J}^\star$). $\mathbf{J}' \subseteq \mathbf{J}$, $\mathbf{J}^\star \subseteq \mathbf{J}$ and $\mathbf{J}' \cap \mathbf{J}^\star = \varnothing$. For each junction point $\mathbf{j}^\star$ in $\mathbf{J}^\star$, all paths $(\mathbf{j}^\star, \mathbf{e})$ are removed where $\mathbf{e}$ is an endpoint in $\mathbf{E}$ and there is another point $\mathbf{e}'$ in $\mathbf{E}$ with $\mathcal{D}(\mathbf{j}', \mathbf{e}') > \mathcal{D}(\mathbf{j}', \mathbf{e})$. $\mathcal{D}$ denotes the distance between two points. The pruned skeleton is shown in Figure 7.11 (b). The endpoints (red dots) and junction points (green dots) of the pruned skeleton are denoted by $\mathbf{E}'$ and $\mathbf{J}'$, respectively. As illustrated in Figure 7.11 (b), for each endpoint $\mathbf{e}_i$, there is one branch $\mathbf{g}_i$ that directly connects to a junction point $\mathbf{j}_m$. $\mathbf{e}_i \in \mathbf{E}'$, $(i = 1, \cdots, N)$ and $\mathbf{j}_m \in \mathbf{J}'$, $(m = 1, \cdots, M)$. Based on the pruned skeleton, for each endpoints $\mathbf{e}_i$ in $\mathbf{E}'$, its geometrical and topological features are described by a six-dimensional vector:

$$\mathbf{e}_i = (L(\mathbf{g}_i), \mathcal{D}_i, l(\mathbf{e}_i), l'(\mathbf{e}_i), \Theta(\mathbf{e}_i), \Theta'(\mathbf{e}_i))^{\mathrm{T}} \quad . \tag{7.2}$$

in which $L(\mathbf{g}_i)$ represents the length of a branch $\mathbf{g}_i$ associated with $\mathbf{e}_i$. $\mathcal{D}_i$ denotes the horizontal distance from $\mathbf{e}_i$ to the leftmost junction point. $l(\mathbf{e}_i)$ and $l'(\mathbf{e}_i)$ denote the mean distances from $\mathbf{e}_i$ to each endpoint and junction point, respectively. Based on the idea of Point Context in Section 4.2, $l(\mathbf{e}_i)$ and $l'(\mathbf{e}_i)$ are calculated by the mean Euclidean distance in the log space (Equation (4.2)). The remaining two features $\Theta(\mathbf{e}_i), \Theta'(\mathbf{e}_i)$ present the mean pairwise orientations of vectors from $\mathbf{e}_i$ to each endpoint and junction point, respectively. These two features are calculated by the four quadrant inverse tangent in Equation (4.3). Eventually, based on the above method, for a given envelope, its shape D is represented by the feature vectors of skeleton endpoints $\mathbf{e}_i$.

Audio Shape Matching: Let $\mathbf{E}_1$ and $\mathbf{E}_2$ be the sets of skeleton endpoints from two shapes D_1 and D_2 respectively. $\mathbf{e}_i$ and $\mathbf{e}'_j$ denote the single endpoints in $\mathbf{E}_1$ and $\mathbf{E}_2$ respectively. Therefore, $\mathbf{e}_i = (L(\mathbf{g}_i), \mathcal{D}_i, l(\mathbf{e}_i), l'(\mathbf{e}_i), \Theta(\mathbf{e}_i), \Theta'(\mathbf{e}_i))^{\mathrm{T}}$ and $\mathbf{e}'_j = (L(\mathbf{g}'_j), \mathcal{D}'_j, l(\mathbf{e}'_j), l'(\mathbf{e}'_j), \Theta(\mathbf{e}'_j), \Theta'(\mathbf{e}'_j))^{\mathrm{T}}$. Let N and M be the numbers of endpoints in $\mathbf{E}'_1$ and $\mathbf{E}'_2$, respectively, where $N \leqslant M$. Here, only the skeleton branches, time series order, as well as endpoint locations are used for searching correspondences due to the following reasons: Firstly, unlike traditional skeletons, Audio Skeletons restrictedly follow the time series order. Secondly, to carry enough audio features, amplitude envelopes normally have more convex than the ordinary object shapes [LLE00]. Consequently, Audio Skeletons have too many endpoints to use the heavy traditional skeleton graph-based matching algorithms [BL08]. Thirdly, using full features to search for correspondences might trigger the overfitting problem. Lastly, the preliminary experiments in [Yan+15c] show that these features are sufficient for investigating promising correspondences. Therefore, in order to fully use the properties of Audio Skeleton and reduce the computational complexity for matching, correspondences between Audio Skeleton endpoints are searched using only the length of associated branch and horizontal distance features.

More specifically, each endpoint set $\mathbf{E}_1$ and $\mathbf{E}_2$ is divided into two groups, up group $\mathbf{E}_1^{(up)}$, $\mathbf{E}_2^{(up)}$ and under group $\mathbf{E}_1^{(under)}$, $\mathbf{E}_2^{(under)}$, based on whether the vertical location of an endpoint is smaller or bigger than the mean vertical values of skeleton junction points, respectively. Consequently, $\mathbf{E}_1^{(up)} = \{\mathbf{e}_1, \cdots, \mathbf{e}_{N'}\}$, $\mathbf{E}_1^{(under)} = \mathbf{E}_1 - \mathbf{E}_1^{(up)}$, $\mathbf{E}_2^{(up)} = \{\mathbf{e}'_1, \cdots, \mathbf{e}'_{M'}\}$, $\mathbf{E}_2^{(under)} = \mathbf{E}_2 - \mathbf{E}_2^{(up)}$. After that, the correspondences of endpoints in two groups are searched independently. As discussed above, skeleton branches and time series order should also be considered for search correspondences. Therefore, the associated branch length and horizontal distance features are considered to calculate the dissimilarity $d(\mathbf{l}_i, \mathbf{l}'_j)$ between $\mathbf{l}_i$ and $\mathbf{l}'_j$ by their Euclidean distance, $\mathbf{l}_i = (L(\mathbf{g}_i), \mathcal{D}_i)^{\mathrm{T}}$ and $\mathbf{l}'_j = (L(\mathbf{g}'_j), \mathcal{D}'_j)^{\mathrm{T}}$. For a skeleton endpoint group $\mathbf{E}_1^{(up)}$ and $\mathbf{E}_2^{(up)}$ from two shapes D_1 and D_2, all dissimilarities between endpoints are computed and then a matrix is obtained:

$$U(\mathbf{E}_1^{(up)}, \mathbf{E}_2^{(up)}) = \begin{pmatrix} d(\mathbf{l}_1, \mathbf{l}'_1) & d(\mathbf{l}_1, \mathbf{l}'_2) & \dots & d(\mathbf{l}_1, \mathbf{l}'_{M'}) \\ d(\mathbf{l}_2, \mathbf{l}'_1) & d(\mathbf{l}_2, \mathbf{l}'_2) & \dots & d(\mathbf{l}_2, \mathbf{l}'_{M'}) \\ \vdots & \vdots & \vdots & \\ d(\mathbf{l}'_{N'}, \mathbf{l}'_1) & d(\mathbf{l}_{N'}, \mathbf{l}'_2) & \dots & d(\mathbf{l}_{N'}, \mathbf{l}'_{M'}) \end{pmatrix} . \tag{7.3}$$

Based on $U(\mathbf{E}_1^{(up)}, \mathbf{E}_2^{(up)})$, the best matched points from $\mathbf{E}_1^{(up)}$ to $\mathbf{E}_2^{(up)}$ are found by the Hungarian algorithm [Kuh55]. For each endpoint $\mathbf{e}_i$ in $\mathbf{E}_1^{(up)}$, the Hungarian algorithm can find its corresponding endpoint $\mathbf{e}'_j$ in $\mathbf{E}_2^{(up)}$. Since $\mathbf{E}_1^{(up)}$ and $\mathbf{E}_2^{(up)}$ may have different numbers of endpoints, the total dissimilarity value should include the penalty for endpoints that cannot find any partner. To achieve this, additional rows are added with a constant value *const* to Equation (7.3) so that $U(\mathbf{E}_1^{(up)}, \mathbf{E}_2^{(up)})$ becomes a square matrix. The constant value *const* is the average of all the other values in $U(\mathbf{E}_1^{(up)}, \mathbf{E}_2^{(up)})$. Analogous to this, for the group $\mathbf{E}_1^{(under)}$ and $\mathbf{E}_2^{(under)}$, the correspondences are also searched by applying the Hungarian algorithm on the dissimilarity matrix $U(\mathbf{E}_1^{(under)}, \mathbf{E}_2^{(under)})$. The final correspondences between $\mathbf{E}_1$ and $\mathbf{E}_2$ are obtained by fusing the correspondences from up and under groups. These correspondences will be used for calculating the similarity between Audio Skeletons using their full features.

Next, the dissimilarity between $\mathbf{E}_1$ and $\mathbf{E}_2$ is calculated by all distance and angle features of their corresponding skeleton endpoints. More specifically, in order to ensure the numerical integration, the original feature vectors of a pair of corresponding endpoints $\mathbf{e}_i$ and $\mathbf{e}'_j$ are firstly divided into distance and angle sub-vectors: $\mathbf{e}_i$ is divided into $\mathbf{e}_{1i} = (L(\mathbf{g}_i), \mathcal{D}_i, l(\mathbf{e}_i), l'(\mathbf{e}_i))^{\mathrm{T}}$, $\mathbf{e}_{2i} = (\Theta(\mathbf{e}_i), \Theta'(\mathbf{e}_i))^{\mathrm{T}}$ and $\mathbf{e}'_j$ is divided into $\mathbf{e}'_{1j} = (L(\mathbf{g}'_j), \mathcal{D}'_j, l(\mathbf{e}'_j), l'(\mathbf{e}'_j))^{\mathrm{T}}$, $\mathbf{e}'_{2j} = (\Theta(\mathbf{e}'_j), \Theta'(\mathbf{e}'_j))^{\mathrm{T}}$. After that, the dissimilarity between $\mathbf{e}_i$ and $\mathbf{e}'_j$ is calculated by:

$$d(\mathbf{e}_i, \mathbf{e}'_j) = d_1(\mathbf{e}_{1i}, \mathbf{e}'_{1j}) + d_2(\mathbf{e}_{2i}, \mathbf{e}'_{2j}) \quad . \tag{7.4}$$

Particularly, inspired by the distance methods in Equation (5.12), $d_1(\mathbf{e}_{1i}, \mathbf{e}'_{1j})$ is calculated by the correlation between distance features $\mathbf{e}_{1i}$ and $\mathbf{e}'_{1j}$:

$$d_1(\mathbf{e}_{1i}, \mathbf{e}'_{1j}) = \frac{\frac{1}{4}\sum_{k=1}^{4}(\mathbf{e}_{1i,k}\mathbf{e}'_{1j,k} - \mu_{\mathbf{e}_{1i}}\mu_{\mathbf{e}'_{1j}})^2}{\sigma_{\mathbf{e}_{1i}}\sigma_{\mathbf{e}'_{1j}}} \quad . \tag{7.5}$$

where μ and σ are the mean and the standard deviation functions, respectively. Inspired by the method in Equation (5.13), $d_2(\mathbf{e}_{2i}, \mathbf{e}'_{2j})$ is calculated by the mean difference between the angle features $\mathbf{e}_{2i}$ and $\mathbf{e}'_{2j}$

$$d_2(\mathbf{e}_{2i}, \mathbf{e}'_{2j}) = \frac{\frac{1}{2}\sum_{k=1}^{2}(\mathbf{e}_{2i,k} - \mathbf{e}'_{2j,k})^2}{\delta^2} \quad . \tag{7.6}$$

where δ represents the tolerance to ensure the numerical integration with $d_1(\mathbf{e}_{1i}, \mathbf{e}'_{1j})$. With the preliminary experiments in [Yan+15c], δ is set to 10. Finally, the global dissimilarity $d(\mathbf{E}_1, \mathbf{E}_2)$ is calculated by the mean dissimilarity among the matched endpoints.

Evaluation: In this part, the AMATI dataset [YGL15] is employed to evaluate the usability of the proposed method on a real application. The AMATI dataset is composed of the recorded violin sounds during the 10th International Henryk Wieniawski Violinmaking Competition in Poznan (2001) and it contains sounds of 52 violins played by the same violinist. The analysis has been performed on a 5s excerpt of J.S. Bach's Partita no 2 in D minor, Sarabande (BWV 1004). Their envelopes are generated (examples in Figure 7.12) with the smoothness coefficient $\kappa = 19$ and iteration $\eta' = 30$.

In the first part of the evaluation, the proposed skeleton pruning method is compared to the DCE-based approach in [BLL07]. As shown in Figure 7.13, it can be clearly observed that the proposed method achieves as good results as the DCE-based method. Moreover, it

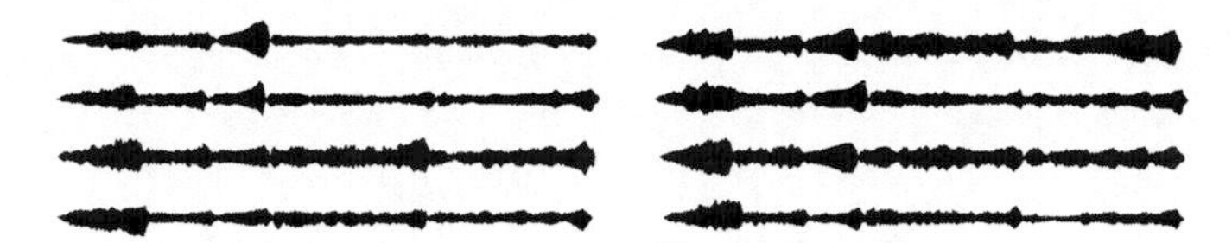

Figure 7.12: Some of the audio envelope shapes in the AMATI.

keeps more skeleton endpoints which are crucial for preserving the audio envelope shape deformation in the fine-grained levels.

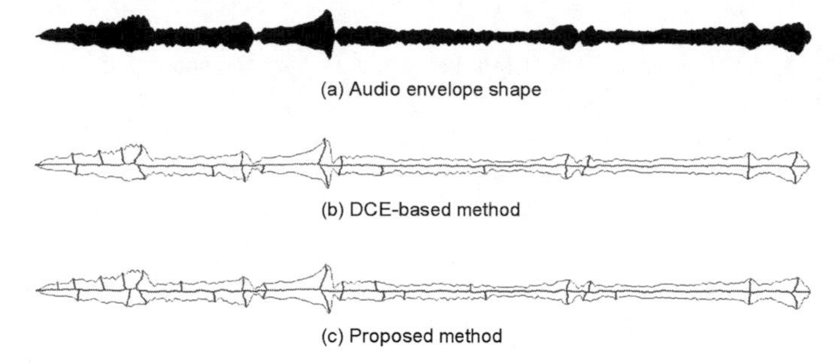

Figure 7.13: Result comparison between the proposed skeleton pruning method and the DCE-based approach [BLL07].

In order to quantitatively evaluate the performance of pruning approaches, two pruning methods are compared in terms of pruning time and the accomplishment of skeleton endpoints. For the comparison of pruning time, a laptop with Inter Core i7 2.2GHz CPU, 8.00GB memory and 64-bit Windows 8.1 OS is employed for the experiment. Since the DCE-based method for skeleton pruning requires a proper stopping parameter to calibrate the pruning power (Section 4.1), it is provided randomly and the mean time is calculated as the pruning time. On the contrary, the proposed method can be applied automatically. For the audio shape (Figure 7.13 (a)) with the size of 93x1490 pixels, the DCE-based method [BLL07] needs 176.23 seconds for generating and pruning the audio skeleton while the proposed method only requires 3.68s. For the comparison of the accomplishment of skeleton endpoints, the pruned skeletons from the DCE-based method are considered as the ground truth and their endpoints are compared to the endpoints in the skeletons which are pruned by the proposed method. The percentage of endpoints which have the same location between two types of skeletons is employed for evaluating the accomplishment. Based on the AMATI dataset, all percentages are collected and the mean value is 94.73%. From the two comparisons above, it can be clearly observed that the proposed skeleton pruning method yields promising results while using much less running time.

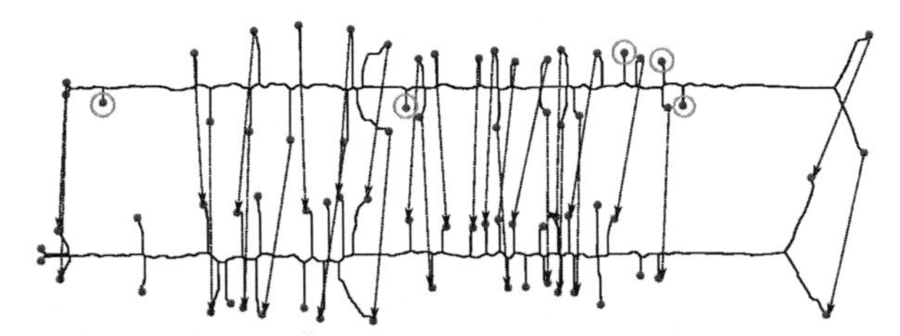

Figure 7.14: The correspondences (marked by dotted arrow lines) between endpoints (marked by red dots) of two audio skeletons.

Figure 7.14 shows the correspondences between the endpoints of two audio skeletons. Since the matching algorithm strictly refers to the skeleton branches and time series order, some endpoints are ignored (marked by green circles in Figure 7.14) if they cannot find the appropriate correspondences in their positions. Comparing the skeleton matching algorithm to [BL08], the proposed algorithm operates much faster since only the Euclidean distance is used on two-dimensional feature vectors for calculating the distance between skeleton endpoints. This character can ensure the proposed algorithm working in real time. Moreover, unlike the global matching in [BL08], the proposed algorithm is applied independently on different endpoint groups which can reduce the searching space and improve the accuracy.

Table 7.2: Audio envelope shape retrieval on AMATI. The bold ID represents the incorrect result based on artificial validation.

query	1	2	3	4	5	6	7	8	9	10	accuracy
10	84	58	**36**	20	40	118	113	**91**	43	**56**	70%
100	109	**93**	108	31	78	**15**	116	32	92	77	80%
20	**36**	**91**	84	58	**40**	113	118	10	35	79	70%
24	46	112	11	**85**	21	**104**	32	17	78	92	80%
32	78	92	**104**	21	46	**11**	112	116	24	**85**	70%
40	118	113	35	**36**	58	**91**	20	41	79	49	80%
74	**56**	**76**	**43**	**72**	**30**	**117**	**84**	**10**	**20**	**36**	0%
78	32	116	92	**104**	21	46	93	109	24	112	90%
85	46	112	24	**11**	17	21	**104**	92	32	80	80%
43	**56**	76	30	10	84	**74**	58	**36**	20	**91**	60%

Finally, based on the AMATI dataset, 10 audio shapes are randomly selected and each of them is used as a query to perform the retrieval on the whole dataset. For each retrieval, the retrieved shapes are ranked according to their similarity $(1 - d(\mathbf{E}_1, \mathbf{E}_2))$ to the query and the top 10 shapes with the highest similarity are selected for analysis. Table 7.2 illustrates the retrieval results among 10 queries. In order to evaluate the results by human perception, two volunteers are employed (one is familiar with the shape matching and another one is familiar with the violin sound evaluation) and asked them to check the retrieved results. Based on their validations, the incorrect results are marked with bold fonts in Table 7.2.

It can be observed that all results are incorrect for query 74. A possible reason is that query instrument 74 was considered to produce the outstanding sound by the jury during the violin competition. Mapping to its correlated audio envelope shape, it is also difficult to find the similar shapes. Moreover, all the similarity values of the retrieved top 10 shapes to query 74 are lower than 0.64, which is smaller than the average value of 0.8 among the remaining queries. Therefore, it is reasonable to propose a threshold ϵ to filter the retrieved results based on their similarity values to the query. Another similar example for this situation is query 43, which was considered to be one of the worst sounds according to the jury's score. Consequently, it is challenging for query 43 to find the similar shapes as well. If the threshold ϵ is set to 0.65, the average accuracy for 10 queries is improved to 77.5%.

7.4 Summary

This chapter first reviews a broad selection of many shape-based applications to demonstrate the power of shape-based methods. After that, two shape-based applications are introduced based on the coarse- and fine-grained shape descriptors. With the proposed coarse-grained shape descriptor in Section 3.3, an EM classification approach is introduced. Experimental results certify the effectiveness and practicability of the proposed automatic EM classification system emphasising the benefits achieved with the shape descriptor in Section 3.3. The classification rate for manually segmented EM shapes increased from 89.7% (with features in [Li+13b]) to 92.5%. The improvement for semi-automatically segmented images is even more convincing - from 66% to 79.5%, respectively. Due to the very promising results, this system can be considered to possess great potential towards the real-world application of EM recognition.

The second application is inspired by the fine-grained shape descriptor in Section 4.3. With the skeleton model, a novel audio representation and matching methods are introduced. With these methods, a shape of audio envelope is represented by an audio skeleton which can preserve the time and amplitude equivalence by its junction points and endpoints. Consequently, the matching of audio envelopes is substituted by searching for the correspondences of skeleton endpoints. Experiments of correspondence matching illustrate the usability of audio skeleton for envelope matching.

Chapter 8

Conclusion and Future Directions

In this chapter, a comprehensive summary of the introduced shape generation, representation and matching algorithms is provided. Moreover, some observations recognised during the experiments and applications are also summarised. Lastly, an outlook of future works is addressed in terms of method improvements and new directions.

8.1 Conclusion

The last half century has seen the development of many biological or physical theories that have explicitly or implicitly involved object shapes and other spatial entities in the real world. Simultaneously, mathematicians and other researchers have studied the properties of object shapes, and, have been stimulated by many areas where object shapes are useful [SP08]. Moreover, computer scientists and engineers have developed numerous algorithms using object shapes. Bringing this knowledge and experience together, this thesis introduces them in an structured way. Moreover, based on the analysis of these algorithms, this thesis also illustrates several novel methods for object shape generation, representation and matching. With the proposed approaches, several applications are addressed. The chapters in this thesis are also organised following this logic.

For shape generation, a general framework of shape contour detection is presented. Based on the detected shape contour parts, an object shape can be generated by properly filling the hollow spaces. In order to improve the accuracy and efficiency of contour detection, 26 CS descriptors and their correlated matching algorithms are evaluated and discussed. With the recommended combination of descriptors and matching algorithms, shape contour parts are detected by a CS-based open curve matching and voting scheme. In Section 6.1, the generated shapes are compared to the ground truth to prove the effectiveness of the proposed method. Moreover, those shapes are also used in the experiments in Section 6.3.1.5 to illustrate the shape matching performances.

For shape representation, there is always a trade-off between accuracy and efficiency. On the one hand, shape should be described as accurately as possible; on the other hand, a shape descriptor should be compact to simplify indexing and retrieval. Keeping this in mind, two types of shape descriptors are introduced in this part. Firstly, the proposed descriptors capture coarse-grained shape features with low computational complexity and then fuse with some rich descriptors [BMP02; BL08] to improve the description power of individuals. Secondly, the proposed descriptors have a high discriminating ability since the fine-grained shape features are captured and preserved. Experiments in Section 6.2 illustrate that the proposed coarse-grained features have a promising description power and can effectively improve the matching accuracy after fusing with rich descriptors. In Section 6.3, experiments on different datasets prove the high performance of the proposed fine-grained features.

For shape matching, the algorithms are designed based on the type and structure of shape descriptors. Specifically, for the coarse-grained descriptors, shape matchings are applied by searching correspondences between contour parts or the distances between shape feature vectors. In order to improve the matching accuracy and flexibility of coarse-grained descriptors, a supervised optimisation strategy is proposed to control the discrimination power of each feature in a feature space. For the fine-grained descriptors, shape matching is more complex since those descriptors normally contain rich feature structures. In addition to the inherent matching strategies, i.e. one-to-one interesting point matching and skeleton graph matching, the idea of high-order graph matching is also considered to improve the matching accuracy of interesting points and hierarchical skeletons. For this, several potential functions are specifically designed for different descriptors. The experiments in Section 6.2 and 6.3 show the impressive robustness of the proposed methods in an object retrieval scenario.

The thesis is concluded by covering a wide spectrum of shape-based applications. In addition to the review of some existing shape-based applications, object shapes are used for EM classification and audio envelop analysis by the proposed coarse- and fine-grained shape descriptors and matching algorithms. The promising experimental results in Section 7.2 and 7.3 illustrate the usability of the proposed methods.

8.2 Future Directions

Although the proposed shape generation, representation and matching methods achieved promising performances in different experiments and applications, there are still several parts which can be improved. (1) In Section 2.3.2.1, since there are many other existing distance measures, they will be employed for evaluating the signature-based CS descriptors. (2) In Section 4.2, contour interesting points will be detected by considering both the overall shape trend and boundary corners to reduce the FN rate as well as the stability and diversity of interesting points. (3) In Section 5.3.1.3, the Monte-Carlo based methods [SCL12] such as Conditional Random Field will be introduced for solving the high-order graph matching problem. (4) In Equation (5.20), a possible improvement is to estimate optimal weights depending on a given pair of hierarchical skeletons using distance metric learning [Xin+03]. In the future, this idea will be implemented and evaluated. (5) To make the EM classification more effective, in Section 7.2, the following two issues should be addressed. First, since there exist many classes of EMs, a larger dataset of microscopic EM images is required for training and testing. Second, although the employed semi-automatic segmentation method leads to a promising results in the experiments, a full-automatic method is still necessary since an automatic EM classification is more useful and convenient in practice. (6) In Section 7.3, more experiments will be carried out to study the relations between the audio shape and the sound qualities. This study can also be employed for the speech emotion analysis using audio skeletons. The main motivation is built on the fact that a dog can recognise a speaker's emotion without understanding the content of what he/she says. Since an audio skeleton only carries the major changes of an audio envelop, it could be used for speech emotion recognition in which the coarse-grained changes of an audio signal are mainly concerned [AKK11].

In the future, three directions will be considered to extend the shape-based research. In the first one, an object skeleton will be directly extracted from natural images. With traditional methods, an object shape is firstly generated, then skeletonisation and skeleton pruning methods are applied to extract an object skeleton. One disadvantage of this process is that if an object shape cannot be properly segmented, the extracted object skeleton cannot fully present the geometry and topology of the original object. Moreover, since a skeleton is only generated with an object shape, the texture and colour information of the original object cannot be preserved in the skeleton model. In contrast, with the proposed idea, the shape generation step can be avoided and more object features such as colour and texture can be used for skeleton extraction. In addition, with the generated skeletons, the object localisation and recognition tasks can be applied in a more straightforward way. Although Shen *et al.* [She+16a] have investigated an approach with the similar purpose, the generated skeletons are suffered by outliers. In such a case, there are still many parts in need of improvement. In the second direction, the deep learning method [LBH15] will be used for shape generation and recognition. With a deep learning framework [Jia+14], an accurate shape generation and recognition could be realised.

For the last direction, a 2D-to-3D skeleton matching algorithm will be designed for non-rigid 2D-to-3D object matching. The last decade has witnessed a tremendous growth in 3D sensing and printing technologies and the availability of large 3D shape datasets brings forth the need to explore and search in 3D shape collections. As ordinary users are not skilled or convenient to model 3D shapes as the query, one typical way is to use 2D-to-3D shape retrieval approach such as sketch-based shape retrieval [Eit+12], etc. However, the underlying problem of multi-modal similarity between a 3D object and its 2D representation is challenging, especially if one desires to deal with non-rigid shapes [Lah+16]. As skeleton models integrate both geometrical and topological features of 2D and 3D objects, it is reasonable to consider it for non-rigid object matching. Particularly, 2D skeletons can be generated from object shapes [BLL07] or natural images [She+16a]. For 3D object collections, their skeletons can be formed depending on different 3D models like meshes [Au+08] and point clouds [Hua+13], etc. With the 2D-to-3D skeleton matching method, both 2D object shapes and nature images can be easily used as the query for 3D object retrieval.

Abbreviations

2D	two-dimensional 147, 150
3D	three-dimensional 28, 150
BPR	Bending Potential Ratio 46
CNNs	Convolutional Neural Networks 9
CS	Contour Segment v, 13–28, 31, 81–93, 141, 143, 153, 155, 157, 158
DA	Discriminate Analysis 60
DCE	Discrete Curve Evolution 35–37, 39, 46, 47, 55–57, 73, 78, 96, 101, 102, 111, 112, 114, 115, 118, 120–122, 136, 137, 153, 154, 156, 158, 159
DNA	Deoxyribonucleic Acid 127
DP	Dynamic Programming 13, 23, 24, 26–28, 86–88, 90–92
DT	Distance Transform 75, 120, 155
DTW	Dynamic Time Warping 13, 23, 24, 26, 86–88, 90–92
EM	Environmental Microorganism vi, 127–129, 131, 139, 142, 143, 156
FFT	Fast Fourier Transform 50
FMM	Fast Marching Method 49, 50, 109
FN	False Negative 108, 143
FP	False Positive 108
HGM	Hyper-Graph Matching 104, 105
HI	Histogram Intersection 14, 23, 26, 60, 86, 87, 89, 90
ID	Inner-Distance 47, 99, 105–108, 116–119, 158
LPF	Low Pass Filter 50–52

MG	Mutual *k*NN Graph 77, 107, 108, 118, 119
mSVM	Multi-class Support Vector Machine 65
PCCS	Extended Shape Context 102, 103, 158
PSSGM	Path Similarity Skeleton Graph Matching 78, 95, 105–108, 111, 112, 115–119, 121–123, 155, 158, 159
RBF	Radial Basis Function 65, 98, 99
RI	Rotation Invariance 84, 157
RNA	Ribonucleic Acid 127
RRWHM	Reweighted Random Walks-based Hyper-Graph Matching 105
SC	Shape Context 21, 40, 44, 47, 79, 101–103, 105–108, 116–119, 123, 158, 159
SI	Scale Invariance 84, 157
SSC	Skeleton-associated Shape Context 47
SVM	Support Vector Machine 40, 64, 65, 98–100, 128
TAE	True Amplitude Envelope 132
TFS	Temporal Fine Structure 131
TI	Translation Invariance 84, 157

Symbols

α	Angle value. 21, 22, 36, 38, 73
β	Angle value. 38
κ	The smoothing power for audio envelops. 132, 136
$\mathbf{u}$	A set of binary elements. 68, 70, 71
$\mathcal{B}$	The horizontal or vertical border of the middle area. 134
$\mathbf{g}$	A skeleton branch. 134–136
O	Computational complexity. 96, 100, 109, 110, 120
a	A correspondence with context-sensitive meaning. 68–74
b	a correspondence with context-sensitive meaning. 68, 70–74
C	Cost of Support Vector Machine. 98, 99
θ	A set of matching cost within high-order graph matching. 68–74
T	Counter or identifier value with context-sensitive meaning. 57, 72–74, 112, 120, 155
A	Area of contour segment, region, circle or hull in two-dimensional (2D). 16, 17, 39, 40, 154
C	Contour segment or curve in 2D. 14, 16–18, 20, 22–26, 29, 30, 36, 96, 153
H	Number of similar parts between two open curves. 24
K	Curvature of a curve point. 17, 18
W	Width with context-sensitive meaning. 111
$\mathbf{c}$	Feature vector of an open curve in 2D. 36, 38, 62, 63
c	A curve or a correspondence with context-sensitive meaning. 29, 30, 68, 70, 71
$\mathbf{f}$	Contour segment descriptor in 2D. 16–24, 26–28, 84–93, 157

h Height with context-sensitive meaning. 17, 38–40, 154

k Index of an element or a parameter with context-sensitive meaning. 46, 47, 55, 56, 62–64, 69, 70, 78, 101, 111–114, 116, 117, 120, 136, 154, 159

w Width with context-sensitive meaning. 17, 18, 38, 40

$\mathcal{D}$ Distance or dissimilarity between two feature vectors or objects. 22–24, 29, 30, 53, 54, 69, 74, 134–136

$d(\cdot)$ Difference between two values with context-sensitive meaning. 23

d Dissimilarity between two objects, shapes or points. 62–64, 72, 77–79, 135, 136, 139

Ψ An element set with context-sensitive meaning. 29, 50

Υ An element set with context-sensitive meaning. 29, 30

$\mathcal{F}$ Feature set or space. 20–22

f Function with contextual meaning. 18, 19, 22, 23, 37–40, 62, 64, 70, 74, 75, 153

$\mathcal{E}$ A function for high-order graph matching. 68, 70, 71

g A power function. 72, 73

o Similarity between two values or features vectors. 74

ϕ A function that returns the max value inside a distance map. 49, 50

$\mathcal{A}$ An affinity vector. 69

I A collection of all sub-problems in high-order graph matching. 70, 71

λ Partiality between two curves or identifier with context-sensitive meaning. 24

e An index with context-sensitive meaning. 75, 155

i Index of an element, set, vector or matrix. 14, 17–22, 29, 30, 36, 48–50, 52–54, 67–71, 73–76, 134–136, 155

j	Index of an element, set, vector or matrix. 22, 23, 53, 62, 67–71, 73–76, 134–136, 155
v	A index with context-sensitive meaning. 70, 71, 75, 76
L	Length of a path. 134–136
l	Length with context-sensitive meaning. 16, 17, 36, 39, 40, 75, 96, 134, 136, 155
$\mathcal{L}$	Boolean function for valuing contour segment (CS) correspondences. 29
V	A distance map. 49, 50, 154
M	Length or size indicator of a set, vector or matrix dimension. 53, 62, 69, 77, 134, 135
m	Identifier with context-sensitive meaning. 20–24, 28–30, 62–64, 109, 110, 134, 153
n	Identifier with context-sensitive meaning. 20, 21, 23, 24, 28–30, 36–38, 55, 56, 62–64, 99, 109–111, 153, 154
N	Length or size indicator of a set, vector or matrix dimension. 14, 16–21, 23, 24, 28–30, 36–38, 48, 52, 53, 62, 63, 65, 74–76, 96, 100, 109, 110, 120, 128, 134, 135, 153
ω	An object class. 65, 128–130
Γ	A tensor for open curve matching. 28–30, 153
δ	A parameter with context-sensitive meaning. 69, 134, 136
η	A weight factor. 75, 77–79, 132, 136
γ	A parameter with context-sensitive meaning. 70, 98, 99
σ	A parameter with context-sensitive meaning. 62–64, 69, 74, 136
ϑ	A window for searching neighbourhoods. 70
Φ	A Pareto optimum. 24
$\mathbf{E}$	A set of endpoints. 133–136, 139
$\mathbf{J}$	A set of junction points. 133, 134
$\mathbf{P}$	A set of elements. 53, 67–73
$\mathbf{Q}$	A set of elements. 53
$\mathbf{e}$	An endpoint in the audio skeleton. 134–136
$\mathbf{j}$	A junction point in the audio skeleton. 134
$\mathbf{q}$	A contour point with context-sensitive meaning. 53

q	An element in a distance transform matrix. 75, 76
$\mathbf{s}$	A sequence of elements. 48, 50–52
P	polygon of an object shape in 2D. 36, 37, 49, 55, 56, 96, 120, 153, 154
ψ	Contribution of an element. 36
$\mathbf{p}$	Point or rather vector in 2D. 11, 14, 16–22, 38, 39, 48, 52–56, 67–71, 154
r	Radii of a tangent disc. 75, 155
E	A region in a shape. 50
ς	A ranking score of hypotheses. 31
D	An object shape in 2D and 3D 36, 38–40, 48, 49, 53–57, 62–64, 67, 69, 72, 75–79, 96, 100, 109, 120, 134, 135, 154
z	An index with context-sensitive meaning. 74, 75, 155
S	A shape skeleton. 55–57, 72–76, 133, 154, 155
$\mathbf{S}$	A matrix of similarity values. 63, 96
U	A matrix of dissimilarity values. 74, 135, 136
τ	Standard derivation of a set of values. 49
R	Temporary value with context-sensitive meaning. 111
s	Temporary value with context-sensitive meaning. 23, 63, 64, 111
Θ	An orientation between two points. 53, 54, 69, 134, 136
ϵ	A threshold for filtering retrieval results. 139
μ	A threshold or a temporary value with context-sensitive meaning. 49–51, 136
t	Threshold or identifier value with context-sensitive meaning. 11, 12, 19, 24, 57, 154
v	A vertex of a polygon. 36, 37, 96, 120, 153
$\mathbf{l}$	A feature vector for representing features of an audio skeleton endpoint. 135
ρ	A weight with context-sensitive meaning 71
$\mathbf{x}$	A reference point. 49, 50
x	xth position in $\mathbf{x}$-axis or a binary value. 14, 17–19, 68, 70–73

y yth position in $\mathbf{y}$-axis. 14, 17–19

ζ A set of constraints. 68

List of Figures

List of Tables

Bibliography

[ACT09] Ghazi AlNaymat, Sanjay Chawla, and Javid Taheri. "SparseDTW: A Novel Approach to Speed Up Dynamic Time Warping". In: *Australasian Data Mining Conference*. 2009, pp. 117–127.

[AE07] M. Athineos and D.P.W. Ellis. "Autoregressive Modeling of Temporal Envelopes". In: *Signal Procssing* 55.11 (2007), pp. 5237–5245.

[AGM14] Pulkit Agrawal, Ross Girshick, and Jitendra Malik. "Analyzing the performance of multilayer neural networks for object recognition". In: *European Conference on Computer Vision*. 2014, pp. 329–344.

[AJ11] M. U. B. Altaf and B. H. Juang. "Audio signal classification with temporal envelopes". In: *IEEE International Conference on Acoustics, Speech and Signal Processing*. 2011, pp. 469–472.

[AKF08] N. Alajlan, M.S. Kamel, and G.H. Freeman. "Geometry-Based Image Retrieval in Binary Image Databases". In: *IEEE Transactions on Pattern Analysis and Machine Intelligence* 30.6 (2008), pp. 1003–1013.

[AKK11] Moataz El Ayadi, Mohamed S. Kamel, and Fakhri Karray. "Survey on speech emotion recognition: Features, classification schemes, and databases". In: *Pattern Recognition* 44.3 (2011), pp. 572–587.

[AL08] M. Awrangjeb and Guojun Lu. "Robust Image Corner Detection Based on the Chord-to-Point Distance Accumulation Technique". In: *IEEE Transactions on Multimedia* 10.6 (2008), pp. 1059–1072.

[Ala+07] Naif Alajlan, Ibrahim El Rube, Mohamed S. Kamel, and George Freeman. "Shape retrieval using triangle-area representation and dynamic space warping". In: *Pattern Recognition* 40.7 (2007), pp. 1911–1920.

[Ama+11] A. Amanatiadis, V. G. Kaburlasos, A. Gasteratos, and S. E. Papadakis. "Evaluation of shape descriptors for shape-based image retrieval". In: *IET Image Processing* 5.5 (2011), pp. 493–499.

[AMO93] Ravindra K. Ahuja, Thomas L. Magnanti, and James B. Orlin. *Network Flows: Theory, Algorithms, and Applications*. Prentice-Hall, Inc., 1993.

[And79] A.M. Andrew. "Another efficient algorithm for convex hulls in two dimensions". In: *Information Processing Letters* 9.5 (1979), pp. 216–219.

[AO04] T. Adamek and N.E. O'Connor. "A multiscale representation method for nonrigid shapes with a single closed contour". In: *IEEE Transactions on Circuits and Systems for Video Technolog* 14.5 (2004), pp. 742–753.

[Arb+90] K. Arbter, W. E. Snyder, H. Burhardt, and G. Hirzinger. "Application of Affine-Invariant Fourier Descriptors to Recognition of 3-D Objects". In: *IEEE Transactions on Pattern Analysis and Machine Intelligence* 12.7 (1990), pp. 640–647.

[AS05] Fabián Arrebola and Francisco Sandoval. "Corner Detection and Curve Segmentation by Multiresolution Chain-code Linking". In: *Pattern Recognition* 38.10 (2005), pp. 1596–1614.

[AT05] C. Asian and S. Tari. "An axis-based representation for recognition". In: *IEEE International Conference on Computer Vision*. Vol. 2. 2005, pp. 1339–1346.

[Au+08] Oscar Kin-Chung Au, Chiew-Lan Tai, Hung-Kuo Chu, Daniel Cohen-Or, and Tong-Yee Lee. "Skeleton Extraction by Mesh Contraction". In: *ACM Transactions on Graphics* 27.3 (2008), pp. 1–10.

[Bai+10] Xiang Bai, Xingwei Yang, L.J. Latecki, Wenyu Liu, and Zhuowen Tu. "Learning Context-Sensitive Shape Similarity by Graph Transduction". In: *IEEE Transactions on Pattern Analysis and Machine Intelligence* 32.5 (2010), pp. 861–874.

[Bai+12] X. Bai, B. Wang, C. Yao, W. Liu, and Z. Tu. "Co-Transduction for Shape Retrieval". In: *IEEE Transactions on Image Processing* 21.5 (2012), pp. 2747–2757.

[Bau+15] Maximilian Baust, Laurent Demaret, Martin Storath, Nassir Navab, and Andreas Weinmann. "Total variation regularization of shape signals". In: *IEEE Conference on Computer Vision and Pattern Recognition*. 2015, pp. 2075–2083.

[BC94] Donald J Berndt and James Clifford. "Using Dynamic Time Warping to Find Patterns in Time Series". In: *KDD Workshop*. 1994, pp. 359–370.

[Ber+95] D. Bernhard, D. D. Leipe, M. L. Sogin, and K. M. Schlegel. "Phylogenetic relationships of the Nassulida within the phylum Ciliophora inferred from the complete small subunit rRNA gene sequences of Furgasonia blochmanni, Obertrumia georgiana, and Pseudomicrothorax dubius". In: *The Journal of Eukaryot Microbiol* 42.2 (1995), pp. 126–133.

[Bha46] A. Bhattacharyya. "On a Measure of Divergence between Two Multinomial Populations". In: *The Indian Journal of Statistics* 7.4 (1946), pp. 401–406.

[BL08] X. Bai and L.J. Latecki. "Path Similarity Skeleton Graph Matching". In: *IEEE Transactions on Pattern Analysis and Machine Intelligence* 30.7 (2008), pp. 1282–1292.

[BLL07] X. Bai, L.J. Latecki, and Wen yu Liu. "Skeleton Pruning by Contour Partitioning with Discrete Curve Evolution". In: *IEEE Transactions on Pattern Analysis and Machine Intelligence* 29.3 (2007), pp. 449–462.

[BLT09] Xiang Bai, Wenyu Liu, and Zhuowen Tu. "Integrating contour and skeleton for shape classification". In: *IEEE International Conference on Computer Vision Workshops*. 2009, pp. 360–367.

[BM12] Saket Bhardwaj and Ajay Mittal. "A Survey on Various Edge Detector Techniques". In: *Procedia Technology* 4 (2012), pp. 220–226.

[BMP02] S. Belongie, J. Malik, and J. Puzicha. "Shape Matching and Object Recognition Using Shape Contexts". In: *PAMI* 24.4 (2002), pp. 509–522.

[BRB01] G. Borgefors, G. Ramella, and G. Sanniti di Baja. "Hierarchical decomposition of multiscale skeletons". In: *IEEE Transactions on Pattern Analysis and Machine Intelligence* 23.11 (2001), pp. 1296–1312.

[Bro+08] Alexander M. Bronstein, Michael M. Bronstein, Alfred M. Bruckstein, and Ron Kimmel. "Analysis of Two-Dimensional Non-Rigid Shapes". In: *International Journal of Computer Vision* 78.1 (2008), pp. 67–88.

[Bro+09] Alexander M. Bronstein, Michael M. Bronstein, Alfred M. Bruckstein, and Ron Kimmel. "Partial Similarity of Objects, or How to Compare a Centaur to a Horse". In: *International Journal of Computer Vision* 84.2 (2009), pp. 163–183.

[BRW14] X. Bai, C. Rao, and X. Wang. "Shape Vocabulary: A Robust and Efficient Shape Representation for Shape Matching". In: *IEEE Transactions on Image Processing* 23.9 (2014), pp. 3935–3949.

[BST15] G. Bertasius, Jianbo Shi, and L. Torresani. "DeepEdge: A multi-scale bifurcated deep network for top-down contour detection". In: *IEEE Conference on Computer Vision and Pattern Recognition*. 2015, pp. 4380–4389.

[BT88] Roger D Boyle and Richard C Thomas. "Computer vision: A first course". In: Blackwell Scientific Publications, Ltd., 1988, pp. 48–50.

[Can86] John Canny. "A Computational Approach to Edge Detection". In: *IEEE Transactions on Pattern Analysis and Machine Intelligence* 8.6 (1986), pp. 679–698.

[Cao+12] WeiGuo Cao, Ping Hu, YuJie Liu, Ming Gong, and Hua Li. "Gaussian-curvature-derived invariants for isometry". In: *Science China Information Sciences* 56.9 (2012), pp. 1–12.

[CB84] R. Chellappa and R. Bagdazian. "Fourier Coding of Image Boundaries". In: *IEEE Transactions on Pattern Analysis and Machine Intelligence* 6.1 (1984), pp. 102–105.

[CD15] Hyung Jin Chang and Y. Demiris. "Unsupervised learning of complex articulated kinematic structures combining motion and skeleton information". In: *IEEE Conference on Computer Vision and Pattern Recognition*. 2015, pp. 3138–3146.

[CFT08] Longbin Chen, R. Feris, and M. Turk. "Efficient partial shape matching using Smith-Waterman algorithm". In: *IEEE Conference on Computer Vision and Pattern Recognition Workshops*. 2008, pp. 1–6.

[Cha07] Olivier Chapelle. "Training a Support Vector Machine in the Primal". In: *Neural Computation* 19.5 (2007), pp. 1155–1178.

[Chu14] Ellen Booth Church. "Why colors and shapes matter". In: *Little Scholastic* (2014).

[CK11] Ming-Ching Chang and Benjamin B. Kimia. "Measuring 3D shape similarity by graph-based matching of the medial scaffolds". In: *Computer Vision and Image Understanding* 115.5 (2011), pp. 707–720.

[CKK15] Houssem Chatbri, Keisuke Kameyama, and Paul Kwan. "A comparative study using contours and skeletons as shape representations for binary image matching". In: *Pattern Recognition Letters* (2015), pages to appear.

[CLS03] W. Choi, K. Lam, and W. Siu. "Extraction of the Euclidean skeleton based on a connectivity criterion". In: *Pattern Recognition* 36.3 (2003), pp. 721–729.

[Coo+92] T.F. Cootes, D.H. Cooper, C.J. Taylor, and J Graham. "Trainable method of parametric shape description". In: *Image and Vision Computing* 10.5 (1992), pp. 289–294.

[CR11] M.F. Caetano and X. Rodet. "Improved estimation of the amplitude envelope of time-domain signals using true envelope cepstral smoothing". In: *IEEE International Conference on Acoustics, Speech and Signal Processing*. 2011, pp. 4244–4247.

[CT13] Sruti Das Choudhury and Tardi Tjahjadi. "Gait recognition based on shape and motion analysis of silhouette contours". In: *Computer Vision and Image Understanding* 117.12 (2013), pp. 1770–1785.

[CW94] F. S. Cohen and Jin-Yinn Wang. "Part I: Modeling image curves using invariant 3-D object curve models-a path to 3-D recognition and shape estimation from image contours". In: *IEEE Transactions on Pattern Analysis and Machine Intelligence* 16.1 (1994), pp. 1–12.

[Dan80] Per-Erik Danielsson. "Euclidean distance mapping". In: *Computer Graphics and image processing* 14.3 (1980), pp. 227–248.

[Dar+06] P. Daras, D. Zarpalas, A. Axenopoulos, D. Tzovaras, and M. G. Strintzis. "Three-Dimensional Shape-Structure Comparison Method for Protein Classification". In: *IEEE/ACM Transactions on Computational Biology and Bioinformatics* 3.3 (2006), pp. 193–207.

[Dav04] E. R. Davies. *Machine Vision: Theory, Algorithms, Practicalities*. San Francisco, CA, USA: Morgan Kaufmann Publishers Inc., 2004.

[DB13] M. Donoser and H. Bischof. "Diffusion Processes for Retrieval Revisited". In: *IEEE Conference on Computer Vision and Pattern Recognition*. 2013, pp. 1320–1327.

[DDS03] P. Dimitrov, James N. Damon, and K. Siddiqi. "Flux invariants for shape". In: *IEEE Conference on Computer Vision and Pattern Recognition*. 2003, pp. 835–841.

[DPS00] P. Dimitrov, C. Phillips, and K. Siddiqi. "Robust and Efficient Skeletal Graphs". In: *IEEE Conference on Computer Vision and Pattern Recognition*. 2000, pp. 417–423.

[DRB10a] M. Donoser, H. Riemenschneider, and H. Bischof. "Linked edges as stable region boundaries". In: *IEEE Conference on Computer Vision and Pattern Recognition*. 2010, pp. 1665–1672.

[DRB10b] Michael Donoser, Hayko Riemenschneider, and Horst Bischof. "Efficient Partial Shape Matching of Outer Contours". In: *Asian Conference on Computer Vision*. 2010, pp. 281–292.

[DT08] M. R. Daliri and Vincent Torre. "Robust symbolic representation for shape recognition and retrieval". In: *Pattern Recognition* 41.5 (2008), pp. 1782–1798.

[Duc+03] A. Duci, A. J. Yezzi, S. K. Mitter, and S. Soatto. "Shape representation via harmonic embedding". In: *IEEE International Conference on Computer Vision*. 2003, pp. 656–662.

[Duc+11] O. Duchenne, F. Bach, In-So Kweon, and J. Ponce. "A Tensor-Based Algorithm for High-Order Graph Matching". In: *IEEE Transactions on Pattern Analysis and Machine Intelligence* 33.12 (2011), pp. 2383–2395.

[Ebi02] Konrad Ebisch. "A Correction to the Douglas-Peucker Line Generalization Algorithm". In: *Computers & Geosciences* 28.8 (2002), pp. 995–997.

[Eit+12] Mathias Eitz, Ronald Richter, Tamy Boubekeur, Kristian Hildebrand, and Marc Alexa. "Sketch-based shape retrieval." In: *ACM Transactions on Graphics* 31.4 (2012), pp. 1–10.

[EKG10] A. Egozi, Y. Keller, and H. Guterman. "Improving Shape Retrieval by Spectral Matching and Meta Similarity". In: *IEEE Transactions on Image Processing* 19.5 (2010), pp. 1319–1327.

[Faw06] Tom Fawcett. "An Introduction to ROC Analysis". In: *Pattern Recognition Letters* 27.8 (2006), pp. 861–874.

[Fei16] Christian Feinen. "Object Representation and Matching Based on Skeletons and Curves". PhD thesis. Logos Verlag Berlin GmbH, Berlin: University of Siegen, 2016.

[FJS09] Vittorio Ferrari, Frederic Jurie, and Cordelia Schmid. "From Images to Shape Models for Object Detection". In: *International Journal of Computer Vision* 87.3 (2009), pp. 284–303.

[FS07] P. F. Felzenszwalb and J.D. Schwartz. "Hierarchical Matching of Deformable Shapes". In: *IEEE Conference on Computer Vision and Pattern Recognition*. 2007, pp. 1–8.

[FTG06] Vittorio Ferrari, Tinne Tuytelaars, and Luc Van Gool. "Object Detection by Contour Segment Networks". In: *European Conference on Computer Vision*. 2006, pp. 14–28.

[Gor+06] L. Gorelick, M. Galun, E. Sharon, R. Basri, and A Brandt. "Shape Representation and Classification Using the Poisson Equation". In: *IEEE Transactions on Pattern Analysis and Machine Intelligence* 28.12 (2006), pp. 1991–2005.

[Grz+13] M. Grzegorzek, C. Li, J. Raskatow, D. Paulus, and N. Vassilieva. "Texture-Based Text Detection in Digital Images with Wavelet Features and Support Vector Machines". In: *International Conference on Computer Recognition Systems*. 2013, pp. 857–866.

[GSL91] S. J. Greenwood, M. L. Sogin, and D. H. Lynn. "Phylogenetic relationships within the class Oligohymenophorea, phylum Ciliophora, inferred from the complete small subunit rRNA gene sequences of Colpidium campylum, Glaucoma chattoni, and Opisthonecta henneguyi". In: *Journal of Molecular Evolution* 33.2 (1991), pp. 163–174.

[GT00] Yu-Hua Gu and Tardi Tjahjadi. "Coarse-to-fine planar object identification using invariant curve features and B-spline modeling". In: *Pattern Recognition* 33.9 (2000), pp. 1411–1422.

[GTC10] R. Gopalan, Pavan Turaga, and Rama Chellappa. "Articulation-invariant Representation of Non-planar Shapes". In: *European Conference on Computer Vision*. 2010, pp. 286–299.

[Har+15] B. Hariharan, P. Arbelaez, R. Girshick, and J. Malik. "Hypercolumns for object segmentation and fine-grained localization". In: *IEEE Conference on Computer Vision and Pattern Recognition*. 2015, pp. 447–456.

[HF07] M. Sabry Hassouna and Aly A. Farag. "A Highly Accurate Solution to the Eikonal Equation on Cartesian Domains". In: *IEEE Transactions on Pattern Analysis and Machine Intelligence* 29.9 (2007), pp. 1563–1574.

[HKR93] Daniel P Huttenlocher, Gregory A Klanderman, and William J Rucklidge. "Comparing images using the Hausdorff distance". In: *IEEE Transactions on Pattern Analysis and Machine Intelligence* 15.9 (1993), pp. 850–863.

[HMK03] Nikolaus Hansen, Sibylle D. Müller, and Petros Koumoutsakos. "Reducing the Time Complexity of the Derandomized Evolution Strategy with Covariance Matrix Adaptation (CMA-ES)". In: *Evolutionary Computation* 11.1 (2003), pp. 1–18.

[HPF14] Jinlong Hu, Xianrong Peng, and Chengyu Fu. "A novel description based on skeleton and contour for shape matching". In: *International Symposium on High-Power Laser Systems and Applications*. 2014, pp. 1–9.

[HS11] Steven Homer and AlanL. Selman. "Introduction to Complexity Theory". In: *Computability and Complexity Theory*. Texts in Computer Science. Springer US, 2011, pp. 75–80.

[HS15] Byung-Woo Hong and Stefano Soatto. "Shape Matching Using Multiscale Integral Invariants". In: *IEEE Transactions on Pattern Analysis and Machine Intelligence* 37.1 (2015), pp. 151–160.

[HS92] Robert M. Haralick and Linda G. Shapiro. *Computer and Robot Vision*. 1st. Boston, MA, USA: Addison-Wesley Longman Publishing Co., Inc., 1992, pp. 346–351.

[HS98] J. W. Harris and H. Stocker. "Segment of a Circle". In: *Handbook of Mathematics and Computational Science*. New York: Springer-Verlag, 1998, pp. 92–93.

[Hua+13] Hui Huang, Shihao Wu, Daniel Cohen-Or, Minglun Gong, Hao Zhang, Guiqing Li, and Baoquan Chen. "L1-medial Skeleton of Point Cloud". In: *ACM Transactions on Graphics* 32.4 (2013), pp. 1–8.

[Huz11] Miljenko Huzak. "International Encyclopedia of Statistical Science". In: 2011. Chap. Chi-Square Distribution, pp. 245–246.

[Ish09] H. Ishikawa. "Higher-order clique reduction in binary graph cut". In: *IEEE Conference on Computer Vision and Pattern Recognition*. 2009, pp. 2993–3000.

[JB14] Faraz Janan and Michael Brady. "Shape Description and Matching Using Integral Invariants on Eccentricity Transformed Images". In: *International Journal of Computer Vision* 113.2 (2014), pp. 92–112.

[Jeg+10] H. Jegou, C. Schmid, H. Harzallah, and J. Verbeek. "Accurate Image Search Using the Contextual Dissimilarity Measure". In: *IEEE Transactions on Pattern Analysis and Machine Intelligence* 32.1 (2010), pp. 2–11.

[Jeg+12] H. Jegou, F. Perronnin, M. Douze, J. Sanchez, P. Perez, and C. Schmid. "Aggregating Local Image Descriptors into Compact Codes". In: *IEEE Transactions on Pattern Analysis and Machine Intelligence* 34.9 (2012), pp. 1704–1716.

[Jia+14] Yangqing Jia, Evan Shelhamer, Jeff Donahue, Sergey Karayev, Jonathan Long, Ross Girshick, Sergio Guadarrama, and Trevor Darrell. "Caffe: Convolutional Architecture for Fast Feature Embedding". In: *ACM International Conference on Multimedia*. 2014, pp. 675–678.

[Kar+15] Pawel Karczmarek, Adam Kiersztyn, Witold Pedrycz, and Przemyslaw Rutka. "Chain Code-Based Local Descriptor for Face Recognition". In: *International Conference on Computer Recognition Systems*. 2015, pp. 307–316.

[KDB10] Peter Kontschieder, Michael Donoser, and Horst Bischof. "Beyond Pairwise Shape Similarity Analysis". In: *Asian Conference on Computer Vision*. 2010, pp. 655–666.

[KGV83] S. Kirkpatrick, C. D. Gelatt, and M. P. Vecchi. "Optimization by simulated annealing". In: *Science* (1983), pp. 671–680.

[Kha+15] N. Khan, M. Algarni, A. Yezzi, and G. Sundaramoorthi. "Shape-tailored local descriptors and their application to segmentation and tracking". In: *IEEE Conference on Computer Vision and Pattern Recognition*. 2015, pp. 3890–3899.

[KK03] E. Keogh and Sh. Kasetty. "On the Need for Time Series Data Mining Benchmarks: A Survey and Empirical Demonstration". In: *Data Mining and Knowledge Discovery* 7.4 (2003), pp. 349–371.

[KLS89] A. Krzyzak, S.Y. Leung, and C.Y. Suen. "Reconstruction of two-dimensional patterns from Fourier descriptors". In: *Machine Vision and Applications* 2.3 (1989), pp. 123–140.

[KP09] N. Komodakis and N. Paragios. "Beyond pairwise energies: Efficient optimization for higher-order MRFs". In: *IEEE Conference on Computer Vision and Pattern Recognition*. 2009, pp. 2985–2992.

[KR07] V. Kolmogorov and C. Rother. "Minimizing Nonsubmodular Functions with Graph Cuts-A Review". In: *IEEE Transactions on Pattern Analysis and Machine Intelligence* 29.7 (2007), pp. 1274–1279.

[KSP95] H. Kauppinen, T. Seppanen, and M. Pietikainen. "An experimental comparison of autoregressive and Fourier-based descriptors in 2D shape classification". In: *IEEE Transactions on Pattern Analysis and Machine Intelligence* 17.2 (1995), pp. 201–207.

[Kuh55] Harold W. Kuhn. "The Hungarian Method for the assignment problem". In: *Naval Research Logistics Quarterly* 2 (1955), pp. 83–97.

[KWH14] J. J. Kivinen, C. K. Williams, and N. Heess. "Visual boundary prediction: A deep neural prediction network and quality dissection". In: *International Conference on Artificial Intelligence and Statistics*. 2014, pp. 512–521.

[Lah+16] Z. Lahner, E. Rodola, F. R. Schmidt, M. M. Bronstein, and D. Cremers. "Efficient Globally Optimal 2D-to-3D Deformable Shape Matching". In: *IEEE Conference on Computer Vision and Pattern Recognition*. 2016, to appear.

[Lat+07] L.J. Latecki, Qiang Wang, S. Koknar-Tezel, and V. Megalooikonomou. "Optimal Subsequence Bijection". In: *IEEE International Conference on Data Mining*. 2007, pp. 565–570.

[LBH15] Yann LeCun, Yoshua Bengio, and Geoffrey Hinton. "Deep learning". In: *Nature* 521.7553 (2015), pp. 436–444.

[LCL11] Jungmin Lee, Minsu Cho, and Kyoung Mu Lee. "Hyper-graph matching via reweighted random walks". In: *IEEE Conference on Computer Vision and Pattern Recognition*. 2011, pp. 1633–1640.

[LG99] T. Liu and D. Geiger. "Approximate tree matching and shape similarity". In: *IEEE International Conference on Computer Vision*. 1999, pp. 456–462.

[LH05] M. Leordeanu and M. Hebert. "A spectral technique for correspondence problems using pairwise constraints". In: *IEEE International Conference on Computer Vision*. Vol. 2. 2005, pp. 1482–1489.

[Li+13a] C. Li, K. Shirahama, J. Czajkowska, M. Grzegorzek, F. Ma, and B. Zhou. "A Multi Stage Approach for Automatic Classification of Environmental Microorganisms". In: *International Conference on Image Processing Computer Vision and Pattern Recognition*. 2013, pp. 364–370.

[Li+13b] C. Li, K. Shirahama, M. Grzegorzek, F. Ma, and B. Zhou. "Classification of Environmental Microorganisms in Microscopic Images using Shape Features and Support Vector Machines". In: *International Conference on Image Processing*. 2013, pp. 2435–2439.

[Lin+15] Liang Lin, Xiaolong Wang, Wei Yang, and Jian-Huang Lai. "Discriminatively trained and-or graph models for object shape detection". In: *IEEE Transactions on Pattern Analysis and Machine Intelligence* 37.5 (2015), pp. 959–972.

[Liu+07] HaiRong Liu, L.J. Latecki, Wenyu Liu, and Xiang Bai. "Visual Curvature". In: *IEEE Conference on Computer Vision and Pattern Recognition*. 2007, pp. 1–8.

[Liu+13] Yebin Liu, J. Gall, C. Stoll, Qionghai Dai, H.-P. Seidel, and C. Theobalt. "Markerless Motion Capture of Multiple Characters Using Multiview Image Segmentation". In: *IEEE Transactions on Pattern Analysis and Machine Intelligence* 35.11 (2013), pp. 2720–2735.

[LJ07] H. Ling and D.W. Jacobs. "Shape Classification Using the Inner-Distance". In: *IEEE Transactions on Pattern Analysis and Machine Intelligence* 29.2 (2007), pp. 286–299.

[LL99] Longin Jan Latecki and Rolf Lakämper. "Convexity Rule for Shape Decomposition Based on Discrete Contour Evolution". In: *Computer Vision and Image Understanding* 73.3 (1999), pp. 441–454.

[LLE00] L. J. Latecki, R. Lakamper, and T. Eckhardt. "Shape descriptors for non-rigid shapes with a single closed contour". In: *IEEE Conference on Computer Vision and Pattern Recognition*. 2000, pp. 424–429.

[LLS92] L. Lam, S.-W. Lee, and C.Y. Suen. "Thinning methodologies-a comprehensive survey". In: *IEEE Transactions on Pattern Analysis and Machine Intelligence* 14.9 (1992), pp. 869–885.

[LLY10] Hairong Liu, Longin Jan Latecki, and Shuicheng Yan. "Robust Clustering as Ensembles of Affinity Relations". In: *Advances in neural information processing systems*. 2010, pp. 1414–1422.

[LS10] L. Liu and D. Shell. "Assessing Optimal Assignment under Uncertainty: An Interval-based Algorithm". In: *Robotics: Science and Systems*. 2010, pp. 936–953.

[Lu+09a] ChengEn Lu, L. Jan Latecki, N. Adluru, Xingwei Yang, and H. Ling. "Shape guided contour grouping with particle filters". In: *International Conference on Computer Vision*. 2009, pp. 2288–2295.

[Lu+09b] ChengEn Lu, L.J. Latecki, N. Adluru, Xingwei Yang, and Haibin Ling. "Shape guided contour grouping with particle filters". In: *IEEE International Conference on Computer Vision*. 2009, pp. 2288–2295.

[LYL10] H. Ling, Xingwei Yang, and LonginJan Latecki. "Balancing Deformability and Discriminability for Shape Matching". In: *European Conference on Computer Vision*. 2010, pp. 411–424.

[Mah+14] Anil Maheshwari, Jorg-RÃ$\frac{1}{4}$diger Sack, Kaveh Shahbaz, and Hamid Zarrabi-Zadeh. "Improved Algorithms for Partial Curve Matching". In: *Algorithmica* 69.3 (2014), pp. 641–657.

[Man+06] S. Manay, D. Cremers, Byung-Woo Hong, A. J. Yezzi, and S. Soatto. "Integral Invariants for Shape Matching". In: *IEEE Transactions on Pattern Analysis and Machine Intelligence* 28.10 (2006), pp. 1602–1618.

[MB03] F. Mokhtarian and Miroslaw Bober. *Curvature Scale Space Representation: Theory, Applications and MPEG-7 Standardization*. Kluwer Academic, 2003.

[MD89] K. V. Mardia and I. L. Dryden. "Shape Distributions for Landmark Data". In: *Advances in Applied Probability* 21.4 (1989), pp. 742–755.

[MG10] Ralph Mitchell and Ji-Dong Gu. *Environmental microbiology*. John Wiley & Sons, 2010.

[MH80] D. Marr and E. Hildreth. "Theory of Edge Detection". In: *Proceedings of the Royal Society of London B: Biological Sciences* 207.1167 (1980), pp. 187–217.

[ML11a] Tianyang Ma and L.J. Latecki. "From partial shape matching through local deformation to robust global shape similarity for object detection". In: *IEEE Conference on Computer Vision and Pattern Recognition*. 2011, pp. 1441–1448.

[ML11b] Tianyang Ma and L.J. Latecki. "From partial shape matching through local deformation to robust global shape similarity for object detection". In: *IEEE Conference on Computer Vision and Pattern Recognition*. 2011, pp. 1441–1448.

[Moo08] B. C. J. Moore. "The role of temporal fine structure processing in pitch perception, masking, and speech perception for normal-hearing and hearing impaired people". In: *J Assoc Res Otolaryngol* 9.4 (2008), pp. 399–406.

[MQR03] Jr. Maurer C.R., Rensheng Qi, and V. Raghavan. "A linear time algorithm for computing exact Euclidean distance transforms of binary images in arbitrary dimensions". In: *IEEE Transactions on Pattern Analysis and Machine Intelligence* 25.2 (2003), pp. 265–270.

[MR94] N. Mayya and V. T. Rajan. "Voronoi diagrams of polygons: A framework for shape representation". In: *IEEE Conference on Computer Vision and Pattern Recognition*. 1994, pp. 638–643.

[MS05] K. Mikolajczyk and C. Schmid. "A performance evaluation of local descriptors". In: *IEEE Transactions on Pattern Analysis and Machine Intelligence* 27.10 (2005), pp. 1615–1630.

[MSJB15] Jarbas Joaci de Mesquita Sa Junior and Andre Ricardo Backes. "Shape classification using line segment statistics". In: *Information Sciences* 305 (2015), pp. 349–356.

[MV06] G. McNeill and S. Vijayakumar. "Hierarchical Procrustes Matching for Shape Retrieval". In: *IEEE Conference on Computer Vision and Pattern Recognition*. 2006, pp. 885–894.

[MYP14] Michael Maire, StellaX. Yu, and Pietro Perona. "Reconstructive Sparse Code Transfer for Contour Detection and Semantic Labeling". In: *Asian Conference on Computer Vision*. 2014, pp. 273–287.

[Ohm+00] Jens-Rainer Ohm, F. Bunjamin, W. Liebsch, B. Makai, K. Mueller, A. Smolic, and D. Zier. "A set of visual feature descriptors and their combination in a low-level description scheme". In: *Signal Processing: Image Communication* 16 (2000), pp. 157–179.

[OK95] R.L. Ogniewicz and O. KÃ$\frac{1}{4}$bler. "Hierarchic Voronoi Skeletons". In: *Pattern Recognition* 28.3 (1995), pp. 343–359.

[PI97] M. Peura and J. Iivarinen. "Efficiency of Simple Shape Descriptors". In: *Third International Workshop on Visual Form*. 1997, pp. 443–451.

[PKB11] Michael Donoser Peter Kontschieder Hayko Riemenschneider and Horst Bischof. "Discriminative Learning of Contour Fragments for Object Detection". In: *British Machine Vision Conference*. 2011, pp. 1–12.

[PRH08] A. Peter, A. Rangarajan, and J. Ho. "Shape Iane rough: Sliding wavelets for indexing and retrieval". In: *IEEE Conference on Computer Vision and Pattern Recognition*. 2008, pp. 1–8.

[PT10] Nadia Payet and Sinisa Todorovic. "From a Set of Shapes to Object Discovery". In: *European Conference on Computer Vision*. 2010, pp. 57–70.

[Qin+13] M. Qinglin, Y. Meng, Y. Zhenya, and F. Haihong. "An Empirical Envelope Estimation Algorithm". In: *International Congress on Image and Signal Processing*. 2013, pp. 1132–1136.

[RDB10a] Hayko Riemenschneider, Michael Donoser, and Horst Bischof. "Using Partial Edge Contour Matches for Efficient Object Category Localization". In: *European Conference on Computer Vision*. 2010, pp. 29–42.

[RDB10b] Hayko Riemenschneider, Michael Donoser, and Horst Bischof. "Using Partial Edge Contour Matches for Efficient Object Category Localization." In: *European Conference on Computer Vision*. 2010, pp. 29–42.

[Ren+13] Zhou Ren, Junsong Yuan, Jingjing Meng, and Zhengyou Zhang. "Robust Part-Based Hand Gesture Recognition Using Kinect Sensor". In: *IEEE Transactions on Multimedia* 15.5 (2013), pp. 1110–1120.

[Ric54] Bellman Richard. "The theory of dynamic programming". In: *Bulletin of the American Mathematical Society* 60.6 (1954), pp. 503–516.

[RK11] Konstantinos A. Raftopoulos and Stefanos D. Kollias. "The Global Local transformation for noise resistant shape representation". In: *Computer Vision and Image Understanding* 115.8 (2011), pp. 1170–1186.

[RN09] Stuart Russell and Peter Norvig. *Artificial Intelligence: A Modern Approach*. 3rd. Prentice Hall Press, 2009.

[RTG00] Yossi Rubner, Carlo Tomasi, and Leonidas J. Guibas. "The Earth Mover's Distance as a Metric for Image Retrieval". In: *International Journal of Computer Vision* 40.2 (2000), pp. 99–121.

[RWP05] Martin Reuter, Franz-Erich Wolter, and Niklas Peinecke. "Laplace-spectra as fingerprints for shape matching". In: *Proceedings of the 2005 ACM symposium on Solid and physical modeling*. 2005, pp. 101–106.

[SBC08] J. Shotton, A. Blake, and R. Cipolla. "Multiscale Categorical Object Recognition Using Contour Fragments". In: *IEEE Transactions on Pattern Analysis and Machine Intelligence* 30.7 (2008), pp. 1270–1281.

[SC00] David McG. Squire and Terry M. Caelli. "Invariance Signatures: Characterizing Contours by Their Departures from Invariance". In: *Computer Vision and Image Understanding* 77.3 (2000), pp. 284–316.

[SC04] Stan Salvador and Philip Chan. "FastDTW: Toward Accurate Dynamic Time Warping in Linear Time and Space". In: *KDD Workshop on Mining Temporal and Sequential Data*. 2004, pp. 70–80.

[SCL12] Yumin Suh, Minsu Cho, and Kyoung Mu Lee. "Graph Matching via Sequential Monte Carlo". In: *European Conference on Computer Vision*. 2012, pp. 624–637.

[SDD12] P. Sondergaard, R. Decorsiere, and T. Dau. "On the relationship between multi-channel envelope and temporal fine structure". In: *Auditory and Audiological Research*. 2012, pp. 1–8.

[Sel80] Peter H Sellers. "The theory and computation of evolutionary distances: Pattern recognition". In: *Journal of Algorithms* 1.4 (1980), pp. 359–373.

[SGS10] K.E.A. van de Sande, T. Gevers, and C.G.M. Snoek. "Evaluating Color Descriptors for Object and Scene Recognition". In: *IEEE Transactions on Pattern Analysis and Machine Intelligence* 32.9 (2010), pp. 1582–1596.

[She+11] W. Shen, Xiang Bai, Rong Hu, Hongyuan Wang, and Longin Jan Latecki. "Skeleton growing and pruning with bending potential ratio". In: *Pattern Recognition* 44.2 (2011), pp. 196–209.

[She+15] Wei Shen, Xinggang Wang, Yan Wang, Xiang Bai, and Zhijiang Zhang. "DeepContour: A deep convolutional feature learned by positive-sharing loss for contour detection". In: *IEEE Conference on Computer Vision and Pattern Recognition*. 2015, pp. 3982–3991.

[She+16a] Wei Shen, Kai Zhao, Yuan Jiang, Yan Wang, Zhijiang Zhang, and Xiang Bai. "Object Skeleton Extraction in Natural Images by Fusing Scale-associated Deep Side Outputs". In: *IEEE Conference on Computer Vision and Pattern Recognition*. 2016, to appear.

[She+16b] Wei Shen, Yuan Jiang, Wenjing Gao, Dan Zeng, and Xinggang Wang. "Shape recognition by bag of skeleton-associated contour parts". In: *Pattern Recognition Letters* (2016).

[Sho+13] Jamie Shotton, Toby Sharp, Alex Kipman, Andrew Fitzgibbon, Mark Finocchio, Andrew Blake, Mat Cook, and Richard Moore. "Real-time Human Pose Recognition in Parts from Single Depth Images". In: *Communications of the ACM* 56.1 (2013), pp. 116–124.

[SK05] T. B. Sebastian and Benjamin B. Kimia. "Curves vs. skeletons in object recognition". In: *Signal Processing* 85.2 (2005), pp. 247–263.

[SKK04] Thomas B. Sebastian, Philip N. Klein, and Benjamin B. Kimia. "Recognition of Shapes by Editing Their Shock Graphs". In: *IEEE Transactions on Pattern Analysis and Machine Intelligence* 26.5 (2004), pp. 550–571.

[SM06] Eitan Sharon and David Mumford. "2d-shape analysis using conformal mapping". In: *International Journal of Computer Vision* 70.1 (2006), pp. 55–75.

[SM99] Bernhard Scholkopft and Klaus-Robert Mullert. "Fisher discriminant analysis with kernels". In: *Neural networks for signal processing IX* 1.1 (1999), p. 1.

[SMW08] T. Syeda-Mahmood and F. Wang. "Shape-Based Retrieval of Heart Sounds for Disease Similarity Detection". In: *European Conference on Computer Vision*. 2008, pp. 568–581.

[SN96] A. Shashua and N. Navab. "Relative affine structure: canonical model for 3D from 2D geometry and applications". In: *IEEE Transactions on Pattern Analysis and Machine Intelligence* 18.9 (1996), pp. 873–883.

[SP08] Kaleem Siddiqi and Stephen Pizer. *Medial representations: mathematics, algorithms and applications*. Vol. 37. Springer Science & Business Media, 2008.

[SP14] B. H. Shekar and Bharathi Pilar. "Shape Representation and Classification through Pattern Spectrum and Local Binary Pattern A Decision Level Fusion Approach". In: *International Conference on Signal and Image Processing*. 2014, pp. 218–224.

[SS05] K.B. Sun and B.J. Super. "Classification of contour shapes using class segment sets". In: *IEEE Conference on Computer Vision and Pattern Recognition*. 2005, pp. 727–733.

[Sup06] B. J. Super. "Retrieval from Shape Databases Using Chance Probability Functions and Fixed Correspondence". In: *International Journal of Pattern Recognition and Artificial Intelligence* 20.8 (2006), pp. 1117–1138.

[SV99] Johan AK Suykens and Joos Vandewalle. "Least squares support vector machine classifiers". In: *Neural processing letters* 9.3 (1999), pp. 293–300.

[SZ03] J. Sivic and A. Zisserman. "Video Google: a text retrieval approach to object matching in videos". In: *IEEE International Conference on Computer Vision*. 2003, pp. 1470–1477.

[SZS10] P. Srinivasan, Qihui Zhu, and Jianbo Shi. "Many-to-one contour matching for describing and discriminating object shape". In: *IEEE Conference on Computer Vision and Pattern Recognition*. 2010, pp. 1673–1680.

[TB97] Quang Minh Tieng and W.W. Boles. "Recognition of 2D object contours using the wavelet transform zero-crossing representation". In: *IEEE Transactions on Pattern Analysis and Machine Intelligence* 19.8 (1997), pp. 910–916.

[TC04] Johan Thureson and Stefan Carlsson. "Appearance Based Qualitative Image Description for Object Class Recognition". In: *European Conference on Computer Vision*. 2004, pp. 518–529.

[Tem+10] A. Temlyakov, B.C. Munsell, J.W. Waggoner, and Song Wang. "Two perceptually motivated strategies for shape classification". In: *IEEE Conference on Computer Vision and Pattern Recognition* (2010), pp. 2289–2296.

[TH04] Andrea Torsello and Edwin R. Hancock. "A Skeletal Measure of 2D Shape Similarity". In: *Computer Vision and Image Understanding* 95.1 (2004), pp. 1–29.

[TKR08] Lorenzo Torresani, Vladimir Kolmogorov, and Carsten Rother. "Feature Correspondence Via Graph Matching: Models and Global Optimization". In: *European Conference on Computer Vision*. 2008, pp. 596–609.

[TNOL13] Duc Thanh Nguyen, Philip O. Ogunbona, and Wanqing Li. "A Novel Shape-based Non-redundant Local Binary Pattern Descriptor for Object Detection". In: *Pattern Recogn.* 46.5 (2013), pp. 1485–1500.

[Tve77] Amos Tversky. "Features of similarity." In: *Psychological review* 84.4 (1977), pp. 327–352.

[TW02] Alexandru Telea and Jarke J. van Wijk. "An Augmented Fast Marching Method for Computing Skeletons and Centerlines". In: *Proceedings of the Symposium on Data Visualisation*. Eurographics Association, 2002, 251–ff.

[Val+13] Richard Anthony Valenzano, Shahab Jabbari Arfaee, Jordan Tyler Thayer, Roni Stern, and Nathan R. Sturtevant. "Using Alternative Suboptimality Bounds in Heuristic Search". In: *International Conference on Automated Planning and Scheduling*. 2013, pp. 233–241.

[VK+11] Oliver Van Kaick, Hao Zhang, Ghassan Hamarneh, and Daniel Cohen-Or. "A survey on shape correspondence". In: 30.6 (2011), pp. 1681–1707.

[VO91] Peter J. Van Otterloo. *A Contour-oriented Approach to Shape Analysis*. Hertfordshire, UK: Prentice Hall International Ltd., 1991.

[VS13] A. Venkitaraman and C.S. Seelamantula. "Temporal Envelope Fit of Transient Audio Signals". In: *IEEE Signal Processing Letters* 20.12 (2013), pp. 1191–1194.

[Wan+11] Jingyan Wang, Yongping Li, Xiang Bai, Ying Zhang, Chao Wang, and Ning Tang. "Learning context-sensitive similarity by shortest path propagation". In: *Pattern Recognition* 44.10 (2011), pp. 2367–2374.

[Wan+12a] J. Wang, Xiang Bai, Xinge You, Wenyu Liu, and Longin Jan Latecki. "Shape matching and classification using height functions". In: *Pattern Recognition Letters* 33.2 (2012), pp. 134–143.

[Wan+12b] Junwei Wang, Xiang Bai, Xinge You, Wenyu Liu, and Longin Jan Latecki. "Shape matching and classification using height functions". In: *Pattern Recognition Letters* 33.2 (2012), pp. 134–143.

[WKL15] Fang Wang, Le Kang, and Yi Li. "Sketch-based 3D shape retrieval using Convolutional Neural Networks". In: *IEEE Conference on Computer Vision and Pattern Recognition*. 2015, pp. 1875–1883.

[WL04] Diedrich Wolter and Longin J. Latecki. "Pacific Rim International Conference on Artificial Intelligence". In: Springer Berlin Heidelberg, 2004. Chap. Shape Matching for Robot Mapping, pp. 693–702.

[WT04] Yue Wang and Eam Khwang Teoh. "A novel 2D shape matching algorithm based on B-spline modeling". In: *International Conference on Image Processing*. Vol. 1. 2004, pp. 409–412.

[XHS08] J. Xie, Pheng-Ann Heng, and Mubarak Shah. "Shape matching and modeling using skeletal context". In: *Pattern Recognition* 41.5 (2008), pp. 1756–1767.

[Xin+03] E. P. Xing, Andrew Y. Ng, Michael I. Jordan, and Stuart Russell. "Distance Metric Learning, With Application To Clustering With Side-Information". In: *Advances in Neural Information Processing Systems*. 2003, pp. 505–512.

[XLT09] C. Xu, J. Liu, and X. Tang. "2D Shape Matching by Contour Flexibility". In: *IEEE Transactions on Pattern Analysis and Machine Intelligence* 31.1 (2009), pp. 180–186.

[Yan+16e] Jianyu Yang, Hongxing Wang, Junsong Yuan, Youfu Li, and Jianyang Liu. "Invariant multi-scale descriptor for shape representation, matching and retrieval". In: *Computer Vision and Image Understanding* 145 (2016), pp. 43–58.

[YIJ08] M. Yang, K. K. Idiyo, and R. Joseph. "A Survey of Shape Feature Extraction Techniques". In: *Pattern Recognition* (2008), pp. 43–90.

[YK8] G.U. Yule and M.G. Kendall. "Partial Correlation". In: *An Introduction to the Theory of Statistic*. 258-270, p. 1968.

[YKTL09] X. Yang, S. Koknar-Tezel, and L.J. Latecki. "Locally constrained diffusion process on locally densified distance spaces with applications to shape retrieval". In: *IEEE Conference on Computer Vision and Pattern Recognition*. 2009, pp. 357–364.

[YLL98] Hee Soo Yang, Sang Uk Lee, and Kyoung Mu Lee. "Recognition of 2D Object Contours Using Starting-Point-Independent Wavelet Coefficient Matching". In: *Journal of Visual Communication and Image Representation* 9.2 (1998), pp. 171–181.

[YPL13] X. Yang, L. Prasad, and L. J. Latecki. "Affinity Learning with Diffusion on Tensor Product Graph". In: *IEEE Transactions on Pattern Analysis and Machine Intelligence* 35.1 (2013), pp. 28–38.

[YT07] Xinge You and Yuan Yan Tang. "Wavelet-Based Approach to Character Skeleton". In: *IEEE Transactions on Image Processing* 16.5 (2007), pp. 1220–1231.

[Yu+10] Bo Yu, Lei Guo, Tianyun Zhao, and Xiaoliang Qian. "A curve matching algorithm based on Freeman Chain Code". In: *IEEE International Conference on Intelligent Computing and Intelligent Systems*. 2010, pp. 669–672.

[YVV95] Ian T Young and Lucas J Van Vliet. "Recursive implementation of the Gaussian filter". In: *Signal processing* 44.2 (1995), pp. 139–151.

[YWB74] Ian T. Young, Joseph E. Walker, and Jack E. Bowie. "An analysis technique for biological shape. I". In: *Information and Control* 25.4 (1974), pp. 357–370.

[Zen+10] Yun Zeng, Chaohui Wang, Yang Wang, Xianfeng Gu, D. Samaras, and N. Paragios. "Dense non-rigid surface registration using high-order graph matching". In: *IEEE Conference on Computer Vision and Pattern Recognition*. 2010, pp. 382–389.

[Zha+07] Xiaohong Zhang, Ming Lei, Dan Yang, Yuzhu Wang, and Litao Ma. "Multiscale curvature product for robust image corner detection in curvature scale space". In: *Pattern Recognition Letters* 28.5 (2007), pp. 545–554.

[Zhu+08] Qihui Zhu, Liming Wang, Yang Wu, and Jianbo Shi. "Contour Context Selection for Object Detection: A Set-to-Set Contour Matching Approach". In: *European Conference on Computer Vision*. 2008, pp. 774–787.

[ZL02] Dengsheng Zhang and Guojun Lu. "A comparative Study of Fourier Descriptors for Shape Representation and Retrieval". In: *Asian Conference on Computer Vision*. 2002, pp. 646–651.

[ZL04] Dengsheng Zhang and Guojun Lu. "Review of shape representation and description techniques". In: *Pattern Recognition* 37.1 (2004), pp. 1–19.

[ZS08] R. Zass and A. Shashua. "Probabilistic graph and hypergraph matching". In: *IEEE Conference on Computer Vision and Pattern Recognition*. 2008, pp. 1–8.

[ZS84] TY Zhang and Ching Y. Suen. "A fast parallel algorithm for thinning digital patterns". In: *Communications of the ACM* 27.3 (1984), pp. 236–239.

[ZT98] Djemel Ziou and Salvatore Tabbone. "Edge Detection Techniques - An Overview". In: *International Journal of Pattern Recognition and Image Analysis* 8 (1998), pp. 537–559.

Own Publications

[Fei+14] Christian Feinen, Cong Yang, Oliver Tiebe, and Marcin Grzegorzek. "Shape Matching Using Point Context and Contour Segments". In: *Asian Conference on Computer Vision (ACCV 2014)*. Springer LNSC, 2014, pp. 95–110.

[Hed+13] Jens Hedrich, Cong Yang, Christian Feinen, Simone Schaefer, Dietrich Paulus, and Marcin Grzegorzek. "Extended Investigations on Skeleton Graph Matching for Object Recognition". In: *International Conference on Computer Recognition Systems (CORES 2013)*. Springer LNCS, 2013, pp. 371–381.

[OYL15] Ioseb Otskheli, Cong Yang, and Ewa Lukasik. "KeyRoad: Intuitive and Predictive Customisation of Touch-Screen Keyboard for People with Hand Disabilities". In: *Creativity in Intellgent Technologies and Data Science (CIT&DS 2015)*. Springer International Publishing, 2015, pp. 848–858.

[Tie+16] Oliver Tiebe, Cong Yang, Muhammad Hassan Khan, Marcin Grzegorzek, and Dominik Scarpin. "Stripes-based Object Matching". In: *Computer and Information Science*. Springer, 2016, pp. 59–72.

[Yan+14a] Cong Yang, Chen Li, Oliver Tiebe, Kimiaki Shirahama, and Marcin Grzegorzek. "Shape-Based Classification of Environmental Microorganisms". In: *International Conference on Pattern Recognition (ICPR 2014)*. IEEE Computer Society, 2014, pp. 3374–3379.

[Yan+14b] Cong Yang, Oliver Tiebe, Pit Pietsch, Christian Feinen, Udo Kelter, and Marcin Grzegorzek. "Shape-based Object Retrieval by Contour Segment Matching". In: *International Conference on Image Processing (ICIP 2014)*. IEEE Computer Society, 2014, pp. 2202–2206.

[Yan+15a] Cong Yang, Christian Feinen, Oliver Tiebe, Kimiaki Shirahama, and Marcin Grzegorzek. "Shape-based Object Matching Using Point Context". In: *International Conference on Multimedia Retrieval (ICMR 2015)*. ACM, 2015, pp. 519–522.

[Yan+15b] Cong Yang, Oliver Tiebe, Pit Pietsch, Christian Feinen, Udo Kelter, and Marcin Grzegorzek. "Shape-based Object Retrieval and Classification with Supervised Optimisation". In: *International Conference on Pattern Recognition Applications and Methods (ICPRAM 2015)*. Springer, 2015, pp. 204–211.

[Yan+15c] Cong Yang, Oliver Tiebe, Marcin Grzegorzek, and Ewa Lukasik. "Skeleton-based Audio Envelope Shape Analysis". In: *Asian Conference on Pattern Recognition (ACPR 2015)*. IEEE Computer Society, 2015, Accepted for Publication.

[Yan+16a] Cong Yang, Oliver Tiebe, Kimiaki Shirahama, Ewa Lukasik, and Marcin Grzegorzek. "Evaluating Contour Segment Descriptors". In: *Machine Vision and Applications* (2016), Under Review.

[Yan+16b] Cong Yang, Oliver Tiebe, Marcin Grzegorzek, and Bipin Indurkhya. "Investigations on Skeleton Completeness for Skeleton-based Shape Matching". In: *Signal Processing: Algorithms, Architectures, Arrangements, and Applications (SPA 2016)*. IEEE Signal Processing Society, 2016, pp. 113–118.

[Yan+16c] Cong Yang, Oliver Tiebe, Kimiaki Shirahama, and Marcin Grzegorzek. "Object Matching with Hierarchical Skeletons". In: *Pattern Recognition* 55 (2016), pp. 183–197.

[Yan+16d] Cong Yang, Christian Feinen, Oliver Tiebe, Kimiaki Shirahama, and Marcin Grzegorzek. "Shape-based object matching using interesting points and high-order graphs". In: *Pattern Recognition Letters* (2016), Accepted for Publication.

[YD12] Cong Yang and Wen long Du. "Research on the Health Diagnosis Module of Large-scale Clusters". In: *International Conference on Computer Science and Network Technology (ICCSNT 2012)*. 2012, pp. 589–593.

[YG14] Cong Yang and Marcin Grzegorzek. "Object Similarity by Humans and Machines". In: *AAAI Fall Symposium on Modeling Changing Perspectives (AAAI MCP)*. AAAI Press, 2014.

[YG15] Cong Yang and Marcin Grzegorzek. "Shape-based Object Matching and Classification". In: *Proceedings of the Joint Workshop of the German Research Training Groups in Computer Science*. Ed. by A. Babari and C. Carapelle. Pro Business Digital Printing Deutschland GmbH, 2015. Chap. 6, pp. 95–96.

[YGL15] Cong Yang, Marcin Grzegorzek, and Ewa Lukasik. "Representing The Evolving Temporal Envelope of Musical Instruments Sounds Using Computer Vision Methods". In: *Signal Processing: Algorithms, Architectures, Arrangements, and Applications (SPA 2015)*. IEEE Signal Processing Society, 2015, pp. 76–80.

[YYW10] Xiaoshang Yang, Cong Yang, and Wenyong Wang. "Statistical analysis-based DCT method for face recognition". In: *International Conference on Information Science and Engineering (ICISE 2010)*. 2010, pp. 5403–5406.

Curriculum Vitae

Personal Data

Last Name, First Name	Yang, Cong
Nationality	Chinese
Date, Place of Birth	10th February 1987, Qingyang, China

Education & Professional Experience

September 1994 - July 1999	Primary School Caotan Primary School and Yima Primary School Qingyang, China
September 1999 - July 2002	Middle School Yima Middle School Qingyang, China
September 2002 - July 2005	Senior High School Qingyang No.1 High School Qingyang, China
September 2006 - July 2010	Student of Software Engineering (Bachelor) School of Software Northeast Normal University, Changchun, China
September 2010 - July 2012	Student of Computer Software and Theory (Master) Ideal Research Institute of Information Technology Northeast Normal University, Changchun, China
May 2011 - May 2012	Research Center of Cloud Computing (Visiting Student) Shenzhen Institutes of Advanced Technology Chinese Academy of Sciences, Shenzhen, China
June 2014 - October 2016	Member of the DFG Research Training Group 1564 Imaging New Modalities University of Siegen, Siegen, Germany
October 2012 - October 2016	PhD Student Research Group for Pattern Recognition University of Siegen, Siegen, Germany
January 2015 - February 2015	Visiting Researcher Institute of Computing Science Poznan University of Technology, Poznan, Poland
October 2015	Visiting Researcher Institute of Philosophy Jagiellonian University, Krakow, Poland

In der Reihe *Studien zur Mustererkennung,*

herausgegeben von

Prof. Dr. Ing Heinricht Niemann und Herrn Prof. Dr. Ing. Elmar Nöth

sind bisher erschienen:

1	Jürgen Haas	Probabilistic Methods in Linguistic Analysis ISBN 978-3-89722-565-7, 2000, 260 S.	40.50 €
2	Manuela Boros	Partielles robustes Parsing spontansprachlicher Dialoge am Beispiel von Zugauskunftdialogen ISBN 978-3-89722-600-5, 2001, 264 S.	40.50 €
3	Stefan Harbeck	Automatische Verfahren zur Sprachdetektion, Landessprachenerkennung und Themendetektion ISBN 978-3-89722-766-8, 2001, 260 S.	40.50 €
4	Julia Fischer	Ein echtzeitfähiges Dialogsystem mit iterativer Ergebnisoptimierung ISBN 978-3-89722-867-2, 2002, 222 S.	40.50 €
5	Ulrike Ahlrichs	Wissensbasierte Szenenexploration auf der Basis erlernter Analysestrategien ISBN 978-3-89722-904-4, 2002, 165 S.	40.50 €
6	Florian Gallwitz	Integrated Stochastic Models for Spontaneous Speech Recognition ISBN 978-3-89722-907-5, 2002, 196 S.	40.50 €
7	Uwe Ohler	Computational Promoter Recognition in Eukaryotic Genomic DNA ISBN 978-3-89722-988-4, 2002, 206 S.	40.50 €
8	Richard Huber	Prosodisch-linguistische Klassifikation von Emotion ISBN 978-3-89722-984-6, 2002, 293 S.	40.50 €

9	Volker Warnke	Integrierte Segmentierung und Klassifikation von Äußerungen und Dialogakten mit heterogenen Wissensquellen ISBN 978-3-8325-0254-6, 2003, 182 S. 40.50 €
10	Michael Reinhold	Robuste, probabilistische, erscheinungsbasierte Objekterkennung ISBN 978-3-8325-0476-2, 2004, 283 S. 40.50 €
11	Matthias Zobel	Optimale Brennweitenwahl für die multiokulare Objektverfolgung ISBN 978-3-8325-0496-0, 2004, 292 S. 40.50 €
12	Bernd Ludwig	Ein konfigurierbares Dialogsystem für Mensch-Maschine-Interaktion in gesprochener Sprache ISBN 978-3-8325-0497-7, 2004, 230 S. 40.50 €
13	Rainer Deventer	Modeling and Control of Static and Dynamic Systems with Bayesian Networks ISBN 978-3-8325-0521-9, 2004, 195 S. 40.50 €
14	Jan Buckow	Multilingual Prosody in Automatic Speech Understanding ISBN 978-3-8325-0581-3, 2004, 164 S. 40.50 €
15	Klaus Donath	Automatische Segmentierung und Analyse von Blutgefäßen ISBN 978-3-8325-0642-1, 2004, 210 S. 40.50 €
16	Axel Walthelm	Sensorbasierte Lokalisations-Algorithmen für mobile Service-Roboter ISBN 978-3-8325-0691-9, 2004, 200 S. 40.50 €
17	Ricarda Dormeyer	Syntaxanalyse auf der Basis der Dependenzgrammatik ISBN 978-3-8325-0723-7, 2004, 200 S. 40.50 €
18	Michael Levit	Spoken Language Understanding without Transcriptions in a Call Center Scenario ISBN 978-3-8325-0930-9, 2005, 249 S. 40.50 €

19	Georg Stemmer	Modeling Variability in Speech Recognition ISBN 978-3-8325-0945-3, 2005, 270 S. 40.50 €
20	Frank Deinzer	Optimale Ansichtenauswahl in der aktiven Objekterkennung ISBN 978-3-8325-1054-1, 2005, 370 S. 40.50 €
21	Radim Chrastek	Automated Retinal Image Analysis for Glaucoma Screening and Vessel Evaluation ISBN 978-3-8325-1191-3, 2006, 233 S. 40.50 €
22	Jochen Schmidt	3-D Reconstruction and Stereo Self-Calibration for Augmented Reality ISBN 978-3-8325-1422-8, 2006, 283 S. 40.50 €
23	Marcin Grzegorzek	Appearance-Based Statistical Object Recognition Including Color and Context Modeling ISBN 978-3-8325-1588-1, 2007, 230 S. 40.50 €
24	Lothar Mischke	Teilautomatisierte Verschlagwortung von in altdeutschen Schriftfonts gesetzten Texten mit Hilfe lernender Verfahren ISBN 978-3-8325-1631-4, 2007, 302 S. 40.50 €
25	Tino Haderlein	Automatic Evaluation of Tracheoesophageal Substitute Voices ISBN 978-3-8325-1769-4, 2007, 238 S. 40.50 €
26	Ingo Scholz	Reconstruction and Modeling of Static and Dynamic Light Fields ISBN 978-3-8325-1963-6, 2008, 254 S. 38.50 €
27	Richard Ottermanns	Mustererkennung, Dimensionsreduktion und statistische Modellierung in der Ökologie – dargestellt am Beispiel der Lebensgemeinschaften grasiger Feldraine in deutschen Agrarlandschaften ISBN 978-3-8325-2032-8, 2008, 499 S. 47.00 €
28	Stefan Steidl	Automatic Classification of Emotion-Related User States in Spontaneous Children's Speech ISBN 978-3-8325-2145-5, 2009, 260 S. 43.50 €

29	Andreas Maier	Speech of Children with Cleft Lip and Palate: Automatic Assessment	
		ISBN 978-3-8325-2144-8, 2009, 220 S.	37.00 €
30	Christian Hacker	Automatic Assessment of Children Speech to Support Language Learning	
		ISBN 978-3-8325-2258-2, 2009, 272 S.	39.00 €
31	Jan-Henning Trustorff	Der Einsatz von Support Vector Machines zur Kreditwürdigkeitsbeurteilung von Unternehmen	
		ISBN 978-3-8325-2375-6, 2009, 260 S.	38.00 €
32	Martin Raab	Real World Approaches for Multilingual and Non-native Speech Recognition	
		ISBN 978-3-8325-2446-3, 2010, 168 S.	44.00 €
33	Michael Wels	Probabilistic Modeling for Segmentation in Magnetic Resonance Images of the Human Brain	
		ISBN 978-3-8325-2631-3, 2010, 148 S.	40.00 €
34	Florian Jäger	Normalization of Magnetic Resonance Images and its Application to the Diagnosis of the Scoliotic Spine	
		ISBN 978-3-8325-2779-2, 2011, 168 S.	36.00 €
35	Viktor Zeißler	Robuste Erkennung der prosodischen Phänomene und der emotionalen Benutzerzustände in einem multimodalen Dialogsystem	
		ISBN 978-3-8325-3167-6, 2012, 380 S.	43.50 €
36	Korbinian Riedhammer	Interactive Approaches to Video Lecture Assessment	
		ISBN 978-3-8325-3235-2, 2012, 164 S.	40.50 €
37	Stefan Wenhardt	Ansichtenauswahl für die 3-D-Rekonstruktion statischer Szenen	
		ISBN 978-3-8325-3465-3, 2013, 218 S.	37.00 €
38	Kerstin Müller	3-D Imaging of the Heart Chambers with C-arm CT	
		ISBN 978-3-8325-3726-5, 2014, 174 S.	46.00 €

39	Chen Li	Content-based Microscopic Image Analysis	
		ISBN 978-3-8325-4253-5, 2016, 196 S.	36.50 €
40	Christian Feinen	Object Representation and Matching Based on Skeletons and Curves	
		ISBN 978-3-8325-4257-3, 2016, 260 S.	50.50 €
41	Juan Rafael Orozco-Arroyave	Analysis of Speech of People with Parkinson's Disease	
		ISBN 978-3-8325-4361-7, 2016, 138 S.	38.00 €